Forschungs-/Entwicklungs-/Innovations-Management

Herausgegeben von
H. D. Bürgel (em.), Stuttgart, Deutschland
D. Grosse, vorm. de Pay, Freiberg, Deutschland
C. Herstatt, Hamburg, Deutschland
H. Koller, Hamburg, Deutschland
M. G. Möhrle, Bremen, Deutschland

Die Reihe stellt aus integrierter Sicht von Betriebswirtschaft und Technik Arbeitsergebnisse auf den Gebieten Forschung, Entwicklung und Innovation vor. Die einzelnen Beiträge sollen dem wissenschaftlichen Fortschritt dienen und die Forderungen der Praxis auf Umsetzbarkeit erfüllen.

Herausgegeben von
Professor Dr. Hans Dietmar Bürgel
(em.), Universität Stuttgart

Professor Dr. Hans Koller
Universität der Bundeswehr Hamburg

Professorin Dr. Diana Grosse,
vorm. de Pay, Technische Universität
Bergakademie Freiberg

Professor Dr. Martin G. Möhrle
Universität Bremen

Professor Dr. Cornelius Herstatt
Technische Universität Hamburg-
Harburg

Verena Nedon

Open Innovation in R&D Departments

An Analysis of Employees' Intention to Exchange Knowledge in OI-Projects

With a foreword by
Univ. Prof. Dr. Cornelius Herstatt

Verena Nedon
Hamburg, Germany

Dissertation Technische Universität Hamburg-Harburg, 2014

Forschungs-/Entwicklungs-/Innovations-Management
ISBN 978-3-658-09584-0   ISBN 978-3-658-09585-7 (eBook)
DOI 10.1007/978-3-658-09585-7

Library of Congress Control Number: 2015936903

Springer Gabler
© Springer Fachmedien Wiesbaden 2015
This work is subject to copyright. All rights are reserved by the Publisher, whether the whole or part of the material is concerned, specifically the rights of translation, reprinting, reuse of illustrations, recitation, broadcasting, reproduction on microfilms or in any other physical way, and transmission or information storage and retrieval, electronic adaptation, computer software, or by similar or dissimilar methodology now known or hereafter developed.
The use of general descriptive names, registered names, trademarks, service marks, etc. in this publication does not imply, even in the absence of a specific statement, that such names are exempt from the relevant protective laws and regulations and therefore free for general use.
The publisher, the authors and the editors are safe to assume that the advice and information in this book are believed to be true and accurate at the date of publication. Neither the publisher nor the authors or the editors give a warranty, express or implied, with respect to the material contained herein or for any errors or omissions that may have been made.

Printed on acid-free paper

Springer Gabler is a brand of Springer Fachmedien Wiesbaden
Springer Fachmedien Wiesbaden is part of Springer Science+Business Media
(www.springer.com)

Foreword

Open innovation (OI) has developed into an important branch of innovation research and relevant topic for practice. To cope with the ever-expanding complexity of R&D, companies increasingly open up their innovation processes and integrate external partners (e.g., customers, universities, suppliers) to accelerate their innovation process and/or facilitate the external use of their innovations. In- and outflows of knowledge are central to the OI-philosophy, indicating that open innovation is linked with knowledge management and especially with knowledge exchange. However, this connection is seldom addressed in the literature.

Verena Nedon bases her research on the legitimated observation that despite the wide range of possible OI-research levels, current empirical studies have a strong focus on the organizational level and most widely neglect the micro-foundation, i.e., employees engaged in open innovation. The rare studies analyzing individuals either focus on members of open source projects and other OI-communities or on lead-users. The present dissertation of Ms. Nedon is, therefore, the first study with clear emphasis on employees working for an OI-embracing company and engaging in OI-projects.

Assuming that most innovations of companies have their starting point in the R&D department, R&D employees play an important role in open innovation. By exchanging their knowledge with external partners, they lay the foundation for collaborative innovations. Consequently, to benefit from open innovation, companies need to know, which factors positively influence R&D employees' willingness and intention to exchange knowledge with external partners in OI-projects. To optimally answer this question, Ms. Nedon adopted an elaborated mixed-method approach. Her findings are based on interviews with R&D managers, a survey amongst R&D employees and follow-up group discussions with scholars and R&D managers, allowing a holistic view on the topic.

The research results linked with the competent interpretation and precise presentation confirm the chosen research approach of Ms. Nedon. Her essential contribution to research lies in the well-grounded discussion, application, and extension of the existing theory in the context of open innovation. Practitioners who are involved in setting up OI-projects receive important guidance for their activities, especially in terms of encouraging R&D employees to exchange knowledge with external partners. Therefore, Ms. Nedon's high-quality research constitutes an important contribution in theoretical as well as practical regards.

Hamburg, February 2015

Univ. Prof. Dr. Cornelius Herstatt

Acknowledgement

Innovation is the engine of every company and "[...] *distinguishes between a leader and a follower"* (Steve Jobs). To be ahead of the competition, companies grasp every opportunity to improve and accelerate their innovation processes – even the assistance of external players. The idea to not solely rely on its own resources and abilities, but to take advantage of the knowledge and brainpower of individuals outside its own boundaries is of major interest for companies as well as for researchers, who analyze this phenomenon since 2003 under the umbrella term "open innovation". Following the OI-concept, it is impossible for a company to have all of the required expertise and suitable knowledge in-house, making knowledge exchange with external sources necessary and valuable.

Fundamentally, a company is the sum of its employees and a project the sum of individual efforts. In the case of OI-projects, the success mainly depends on the efforts of a company's R&D employees. By exchanging their knowledge with external partners in OI-projects, they lay the foundation for open innovation. This implies, on the other hand, that their behavior can be a major risk and barrier to open innovation, e.g., if the Not-Invented-Here (NIH) syndrome hampers the acceptance of external knowledge.

Despite the fundamental role of individuals, open innovation has been analyzed mainly on the organizational level, leaving a lot of blank spots on the micro-foundation of this phenomenon. This dissertation views open innovation from the perspective of R&D employees with OI-experience and tries to make a contribution by analyzing why R&D employees engage in knowledge exchange with external partners in OI-projects. The study aims to arouse the attention of researchers as well as managers, interested in the micro-level (i.e., the people side) of open innovation.

This dissertation would not have been possible without the ongoing help of a number of supporters, which I would like to acknowledge and thank for. Without claiming that this list is exhaustive, I especially thank

- Prof. Dr. Cornelius Herstatt for integrating me into his outstanding team and interpreting his role of my "Doktorvater" so literally, as I could not have wished for a more caring and supportive supervisor;
- Prof. Dr. Kathrin M. Möslein for assuming the role of my second evaluator and continuously providing me with valuable feedback;
- Prof. Dr. Christian M. Ringle for assuming the chairman's role of the examination committee;
- Prof. Dr. Andreas Suchanek for supervising my first scientific work on innovation and bringing me into contact with the first company participating in my empirical study;

- All R&D-managers and employees participating in interviews, the online-survey, and follow-up group discussions for taking the time to share their thoughts, provide me with precious insights and, thus, enable the empirical part of my study;
- The outstanding TIM-team for all the inspiring and constructive discussions and all off-topic moments, which turned the time at TIM into an incredible fun time;
- The participants of the XXV ISPIM Conference and several doctoral seminars for all their valuable input and challenging questions;
- My parents for their unconditional love and support. In endless gratefulness, I dedicate this work to you.

Most of all, I thank Konstantin for sharing my dreams, bringing out the best in me, and being the most loyal and warm-hearted companion on a journey that has only just begun.

Munich, February 2015

Verena Nedon

Table of Contents

List of Tables XII
List of Figures XIII
List of Abbreviations and Symbols XIV

1 Introduction 1
    1.1   Research Motivation and Objective 2
    1.2   Research Approach and Contribution 3
    1.3   Structure of Dissertation 5

2 Conceptual Foundation 7
    2.1   Open Innovation 7
        2.1.1   From Closed Innovation to Open Innovation 7
        2.1.2   Prior Research with Focus on External Innovation Sources 10
        2.1.3   Current Developments in OI-Research 13
    2.2   Knowledge Management 30
        2.2.1   Perspectives on Knowledge 30
        2.2.2   Elements of Knowledge Management 32
        2.2.3   Knowledge Exchange 33
    2.3   Research Gap and Derivation of Research Questions 39
    2.4   Chapter Summary 41

3 Theoretical Foundation 43
    3.1   Theory of Planned Behavior 43
        3.1.1   Attitude 45
        3.1.2   Subjective Norm 45
        3.1.3   Perceived Behavioral Control 46
        3.1.4   Intention 47
        3.1.5   Behavior 48
    3.2   TPB and Knowledge Exchange – A Literature Review 49
    3.3   Hypotheses and Research Model 52
        3.3.1   Theory of Planned Behavior 52
        3.3.2   Enjoyment in Helping 53
        3.3.3   Sense of Self-Worth 54

		3.3.4	Reciprocity 55
		3.3.5	Rewards 56
	3.4	Chapter Summary 58

4 Research Design and Operationalization 61
    4.1   Research Approach 61
    4.2   Selection of Companies 63
    4.3   Qualitative Pre-Study (Interviews) 64
           4.3.1   Data Collection and Sample 65
           4.3.2   Method of Data Analysis 66
    4.4   Quantitative Study (Online Survey) 67
           4.4.1   Operationalization of Constructs 67
           4.4.2   Pre-test 76
           4.4.3   Data Collection 77
           4.4.4   Data Cleansing, Data Preparation and Final Sample 77
           4.4.5   Method of Data Analysis 80
    4.5   Chapter Summary 83

5 Findings from Qualitative Pre-Study (Interviews) 85
    5.1   Open Innovation from an R&D Perspective 85
    5.2   Setting up OI-Projects 87
    5.3   Searching and Choosing OI-Partners 89
    5.4   Basic Conditions for OI-Projects 92
    5.5   Benefits and Challenges of Open Innovation 96
    5.6   Chapter Summary 99

6 Findings from Quantitative Study (Online Survey) 101
    6.1   Data Distribution and Bias Treatment 101
           6.1.1   Data Distribution 101
           6.1.2   Bias Treatment 102
    6.2   Descriptive Results 103
    6.3   Findings from an Open-Ended Question 109
    6.4   Measurement Model 112
           6.4.1   Reflective Constructs 112
           6.4.2   Formative Constructs 126
    6.5   Structural Model 128
           6.5.1   Evaluation of Structural Model and Hypotheses 129

		6.5.2	Evaluation of Control Variables	132

6.6 Chapter Summary 135

7 Discussion 137
7.1 RQ1: R&D Perspective on Open Innovation 137
7.2 RQ2: Determinants of R&D Employees' Intention to Exchange Knowledge in OI-Projects 142
7.3 RQ3: Motivational Factors with Positive Influence on R&D Employees' Willingness to Exchange Knowledge in OI-Projects 144

8 Conclusions 149
8.1 Contribution to Academic Research 149
    8.1.1 Contribution to Open Innovation Research and the TBP 149
    8.1.2 Contribution to Knowledge Management Research 151
    8.1.3 Contribution to Motivation Theory 151
8.2 Managerial Implications 152
    8.2.1 Recommendations Related to Attitude ("Want") 154
    8.2.2 Recommendations Related to Subjective Norm ("Shall") 155
    8.2.3 Recommendations Related to Perceived Behavioral Control ("Can") 156
8.3 Limitations and Suggestions for Further Research 157

References 161

Appendix 189

List of Tables

Table 1: Literature Review 50
Table 2: Articles with Predictors of Attitude 51
Table 3: Overview of Interviews 65
Table 4: Operationalization of Theory of Planned Behavior Constructs 68
Table 5: Operationalization of Motivational Constructs 69
Table 6: Decision Rules – Formative versus Reflective 71
Table 7: Sample, Firms, and Responses 80
Table 8: Sample and Sub-Sample Characteristics 104
Table 9: MSA, Communalities and Pattern Matrix – Overall EFA 116
Table 10: MSA, Communalities and Pattern Matrix – All Reflective Constructs (Separately) 118
Table 11: MSA, Communalities and Pattern Matrix – Reward Construct 119
Table 12: MSA, Communalities and Pattern Matrix – Big Five Construct 120
Table 13: EFA Results, ITC, and IIC after Item Exclusion 121
Table 14: Indicator and Internal Consistency Reliability, Convergent Validity 124
Table 15: Correlations and Discriminant Validity 125
Table 16: Cross-Loadings 125
Table 17: Evaluation of Formative Measures of Subjective Norm 127
Table 18: Evaluation of Structural Model 131
Table 19: Evaluation of Hypotheses 132
Table 20: Comparison of Structural Model with and without Control Variables 133
Table 21: Evaluation of Structural Model with Control Variables Included 134

List of Figures

Figure 1: Closed Innovation Model 8
Figure 2: Open Innovation Model 10
Figure 3: Placement of Open Innovation Research 13
Figure 4: Levels of Analysis in Open Innovation Research 14
Figure 5: Macro- and Micro-Level Proposition 16
Figure 6: Phases of (Inbound) Open Innovation 26
Figure 7: Knowledge Management Process 33
Figure 8: Concept of Knowledge Exchange 34
Figure 9: Research Focus 42
Figure 10: Theory of Planned Behavior 44
Figure 11: Theory of Planned Behavior and Sources of Behavioral Barriers 44
Figure 12: Research Model 58
Figure 13: Purposes for Mixed-Method Approach 62
Figure 14: Overview Data Cleansing 78
Figure 15: Advantages and Disadvantages of Open Innovation 100
Figure 16: OI-Partners (OI-Experience) 104
Figure 17: Descriptive Results regarding Intention 106
Figure 18: Descriptive Results regarding Attitude 106
Figure 19: Descriptive Results regarding Subjective Norm 107
Figure 20: Descriptive Results regarding Rewards 108
Figure 21: Descriptive Results regarding Sense of Self-Worth 108
Figure 22: Research Approach for Open-Ended Survey Question 109
Figure 23: Categories of Requirement for Knowledge Exchange in OI-Projects 111
Figure 24: Requirements for Knowledge Exchange in OI-Projects 112
Figure 25: Results from PLS Analysis 130
Figure 26: Relevant Aspects for Knowledge Exchange in OI-Projects 141
Figure 27: Herzberg's Motivation-Hygiene Theory and Reward Constructs ... 146
Figure 28: Recommendations for Managerial Practice along TPB Components 153

List of Abbreviations and Symbols

A	Attitude
AMOS	Analysis of Moment Structures
AVE	Average Variance Extracted
b	Standardized Path Weight/Loading (Path Coefficient)
b_i	Behavioral Belief Strength
B2B	Business-to-Business
B2C	Business-to-Consumer
c_i	Control Belief Strength
CEO	Chief Executive Officer
cf.	Compare
CFA	Confirmatory Factor Analysis
EFA	Exploratory Factor Analysis
e_i	Outcome Evaluation
e.g.	For Example
et al.	And Others
etc.	Et Cetera
f.	And the Following Page
ff.	And the Following Pages
f^2	Effect Size
H	Hypothesis
I	Intention
i.e.	That is
IIC	Inter-Item-Correlation
incl.	Including
IP	Intellectual Property
IT	Information Technology
ITC	Corrected Item-Total-Correlation
JOY	Enjoyment in Helping
KMO	Kaiser-Meyer-Olkin (Criterion)
LISREL	Linear Structural Relationships
m_i	Motivation to Comply
MAR	Missing at Random
MCAR	Missing Completely at Random
MNAR	Missing not at Random
MSA	Measure of Sampling Adequacy
N	Sample Size

n_i	Normative Belief Strength
NDA	Non-Disclosure Agreement
NIH	Not-Invented-Here
NSH	Not-Sold-Here
n.s.	Not Significant
OD	Omission Distance
OI	Open Innovation
p	p-Value
p.	Page
p_i	Control Belief Power
PBC	Perceived Behavioral Control
PLS	Partial Least Squares
pp.	Pages
q^2	Degree of Predictive Relevance
Q^2	Predictive Relevance (Stone-Geisser's Q^2)
REW	Reward
RP	Reciprocity
RQ	Research Question
R&D	Research and Development
R^2	Coefficient of Determination (Explained Variance)
SME	Small and Medium-Sized Enterprises
SN	Subjective Norm
SSL	Sum of Squared Loadings
SW	Sense of Self-Worth
TACT	Target, Action, Context, Time
TPB	Theory of Planned Behavior
TRA	Theory of Reasoned Action
VIF	Variance Inflation Factor
α	Standardized Cronbach's Alpha
λ	Standardized Indicator Loading
ρ	Composite Reliability (Goldstein-Dillon's ρ)

1 Introduction

Companies increasingly face a level of complexity and multi-disciplinarity in their research and development (R&D) of products, which a single player is unable to cope with – especially if he wants to stay competitive (see Miotti and Sachwald 2003; Pfeffer and Salancik 2009). A company can address this issue by opening up its innovation process and integrating external partners and sources (e.g., customers, universities, suppliers) to accelerate its own innovation process and/or facilitate the external use of its innovations (see Chesbrough 2003; Chesbrough et al. 2006). This phenomenon is called open innovation (OI).

> "At its root, Open Innovation assumes that useful knowledge is widely distributed, and that even the most capable R&D organizations must identify, connect to, and leverage external knowledge sources as a core process in innovation." (Chesbrough 2006c, p. 2)

Consequently, relying only on its own resources and abilities is no longer a sustainable option for an innovative company (cf. Caloghirou et al. 2004, p. 31; Fichter 2005, pp. 240ff.). According to a study conducted by the Fraunhofer Institute in collaboration with Henry Chesbrough (the originator of the OI-concept), open innovation has become relevant in various industries (cf. Chesbrough and Brunswicker 2013, p. 6). The motives for engaging in OI-activities are manifold and include inter alia the access to unique knowledge, the exploration of new trends and business opportunities, the mitigation of risks, and the improvement in efficiency, leading to faster time to market (cf. Chesbrough and Brunswicker 2013, p. 18; Fichter 2005, pp. 241ff.; Wallin and Krogh 2010, p. 147). The underlying objective of a company's engagement in open innovation will affect its choice of OI-partner and how the company opens up its innovation process. According to Gassmann and Enkel (2004), a decision can be taken to opt for outside-in OI (i.e., to obtain external knowledge and integrate it in the internal innovation process), or inside-out OI (i.e., the exploitation of internal ideas and technologies outside the company), or a coupled OI-approach (i.e., the combination of outside-in and inside-out activities). Each approach offers different configuration possibilities, so that companies can choose between various options to engage in open innovation (cf. Chesbrough and Brunswicker 2013, p. 10): They can, for instance, enter R&D alliances with different partners, engage in customer co-creation, and/or use crowdsourcing (outside-in OI). They could also set up a spin-off, out-license their IP, and/or enter a joint-venture (inside-out OI). This diversity gives ample scope for the configuration of an individual and company-specific OI-roadmap. To optimally exploit this great potential, most companies adopting open innovation decide for a coupled OI-approach (see Lichtenthaler 2008; Schroll and Mild 2011; Vrande et al. 2009).

1.1 Research Motivation and Objective

Open innovation has not only become a relevant topic for companies, but also for researchers. During the last decade, open innovation has gradually developed into a broad research field with many different streams and various connections to other research areas.[1] The resulting span of OI-research means there are still a lot of blank spaces, even though numerous scholars have already made their contribution to this field.

The inflows and outflows of knowledge are central to the OI-definition (see Chesbrough 2006c), indicating that open innovation is associated with the management of knowledge and especially with the exchange of knowledge. However, this connection is seldom addressed in the literature. Notwithstanding this shortfall, a major gap in OI-research results from the unbalanced examination of different examination objects. Despite the wide range of possible OI-research levels (cf. Vanhaverbeke and Cloodt 2006, pp. 276ff.); current studies have a clear emphasis on the organizational level (cf. West et al. 2006, p. 287). Very few studies focus on individuals (cf. Vrande et al. 2010, p. 226) and employees' perspectives on open innovation are the most widely neglected. The rare studies dealing with open innovation in connection with employees examine OI-relevant competencies and characteristics (see Enkel 2010; Du Chatenier et al. 2010; Pedrosa et al. 2013) or discuss possible individual-related OI-barriers (cf. Enkel 2009, pp. 189ff.), which can be subdivided into three stereotypes: "want-barrier", "shall-barrier", and "can-barrier". Recognizing this imbalance and the fact that a micro-foundation is essential for reliable explanations on a more aggregated level (see Coleman 1990; Felin and Foss 2005), scholars have tried to encourage other researchers to focus more on the level of the individual (cf. Elmquist et al. 2009, pp. 339ff.; Vanhaverbeke 2006, pp. 206f.; Vanhaverbeke and Cloodt 2006, p. 279; Vrande et al. 2010, p. 230; West et al. 2006, pp. 287ff.).

> *"[U]nderstanding the fundamental cogs and wheels of what happens in organizations requires beginning from their fundamental constituents, namely individuals [...]." (Foss et al. 2010, p. 457)*

Employees deserve special attention because they are the ultimate decision-makers in an organization – even though they do not act in a social "vacuum" (cf. Husted and Michailova 2010, p. 40).

> *"[K]nowledge resides within [...] the employees who create, recognize, archive, access, and apply knowledge in carrying out their tasks. Consequently, the movement of knowledge across individual and organizational boundaries, into and from repositories, and into organizational routines and practices is ultimately dependent on employees' knowledge-sharing behaviors." (Bock et al. 2005, p. 88)*

---

[1] For an overview of existing OI-literature, see Dahlander and Gann 2010; Elmquist et al. 2009; Gassmann 2006; Gassmann et al. 2010; Lichtenthaler 2011; Vrande et al. 2010.

Assuming that the R&D department of a company is the place where most companies' innovations begin, R&D employees play an important role in open innovation and so were selected as the object of this study. By exchanging their knowledge with external partners in OI-projects, R&D employees lay the foundation for collaborative innovation. However, this also implies their behavior can be a major risk to open innovation, e.g., if the Not-Invented-Here (NIH) syndrome (see Clagett 1967, Katz and Allen 1982) hampers the acceptance of external knowledge.

> *"Of course, organizational barriers to user solution data do not necessarily end even after the information enters the firm. A firm's R & D group, for example, may well regard such information with a dubious eye. And, given typical incentives and staffing patterns such a reaction, too, is perfectly logical. Note that R & D groups are often staffed with people who are trained to develop new products and processes in-house and are rewarded for this task." (Hippel 1988, p. 119)*

Consequently, companies following an OI-approach do not only depend on co-operation from external partners, but particularly on the support of their R&D employees. To benefit from the OI-approach, companies therefore need to understand their R&D employees' motives for exchanging knowledge in OI-projects. However, very little is known about open innovation at the level of employees and especially about determinants of their knowledge exchange behavior in OI-projects. Therefore, the main objective of my study is to understand why R&D employees become active in OI-projects and why they participate in knowledge exchange with external partners in OI-projects, respectively. Furthermore, I will strive to identify basic conditions facilitating this exchange and to derive implications for companies.

1.2 Research Approach and Contribution

Guided by the aspiration to shed light on the reasoning behind R&D employees' participation in OI-projects in the form of their knowledge exchange with external partners, I have formulated three research questions. To answer them, I have searched the literature for suitable theories that would provide a proper theoretical foundation for my research. In the course of this search, I realized that the three stereotypes of OI-barriers ("want", "shall", "can") influencing individuals' behavior can be related to the theory of planned behavior (TPB). The TPB assumes individuals' intention to behave in a certain manner is determined by three factors: their attitude toward the behavior (associated with "want-barrier"), the subjective norm, or perceived social pressure to perform or not perform the behavior (associated with "shall-barrier"), and the perceived behavioral control (associated with "can-barrier") (see Ajzen 1991). Since the knowledge exchange behavior of individuals had not yet been researched in the context of open innovation, a literature review of publications connecting the TPB and individuals' knowledge exchange behavior was conducted. The goal was to identify motivational factors and related theories that have an impact on employees' willingness to exchange knowledge in OI-projects. Based on the TPB and this literature review, I derived a research model and related hypothesis.

To answer the three research questions optimally, I decided to combine qualitative and quantitative methods (cf. DeCuir-Gunby 2008, pp. 125f.; Walliman 2006, pp. 36f.) and acquired four companies willing to participate in my study. The companies all publicly pronounced the application of open innovation and were manufacturers with global business, headquartered in Germany, and operating in the fields of chemistry, automation, and steel treatment. As a first step, I conducted interviews with 12 OI-experienced R&D managers to understand the R&D perspective on open innovation. Secondly, I initiated an online survey among OI-experienced R&D employees. By means of the resulting 133 usable responses, the research model and related hypotheses could be tested. Lastly, three follow-up group discussions helped to interpret the results from the survey. The application of this empirical mixed method approach allowed me to develop the survey with the help of the interviews; to confirm results from the different methods; and to elaborate and clarify results of the online survey with results from the interviews and group-discussions (cf. Greene et al. 1989, p. 259).

To date, very few empirical studies have connected open innovation with knowledge exchange, examined open innovation in an R&D context, or focused on the personnel's views on open innovation. Furthermore, to my knowledge no study has ever combined these aspects. My thesis is the first empirical study with a clear focus on OI-experienced R&D employees and on the determinants of their intention to exchange knowledge with external partners in OI-projects. The thesis targets a set of relevant questions related to the human side of open innovation and thereby applies the TPB for the first time in an OI-context. It challenges the prior dominant position of the organizational level in OI-studies and, thus, significantly contributes to the micro-foundation of OI-research and to the current understanding of open innovation. The findings give critical insights into open innovation at the level of R&D employees. In detail, I can show that the surveyed R&D employees' attitude is not the dominant determinant of their intention to exchange knowledge with external partners in OI-projects. Instead, the perceived social pressure to exchange knowledge in OI-projects has by far the most powerful impact. The study also reveals the importance of differentiating between the exchange of documented and undocumented knowledge in the context of open innovation. Furthermore, the results show that most of the motivational factors derived from the knowledge management literature help to explain employees' attitude toward their knowledge exchange in OI-projects and that the surveyed R&D employees are mainly intrinsically motivated. Organizational rewards do not have a significant influence on their attitude, but rewards connected to their personal development play a role. This implies that it is worthwhile having a closer look at the reward construct in the context of knowledge exchange in OI-projects and to distinguish among different kinds of rewards. Lastly, my study uncovers that – from an R&D employee's perspective – the most important requirements for participating in knowledge exchange with external partners are related to legal security, a trustful relationship with the external partner, and common ground and fairness between the parties.

The findings of my thesis are not only relevant to the research community that can relate to my results, but also for OI-experienced companies and OI-newcomers. The results indicate how to leverage R&D employees' intention to exchange their knowledge in OI-projects. Furthermore, companies can use the results to reconsider their incentive systems and to reflect if, and to what extent, the general framework of their OI-projects meets the requirements for knowledge exchange in OI-projects.

1.3 Structure of Dissertation

This thesis is structured into eight chapters, with this introduction being *chapter 1*.

In *chapter 2*, the underlying concepts of this study are outlined and research questions are framed, based on the identified research gap. Since the OI-approach is the fundamental concept of this study, antecedents and basic principles are introduced and an overview of prior and current research streams related to open innovation is provided. Furthermore, the link between open innovation and knowledge management is emphasized and OI-relevant aspects of knowledge management are discussed. Finally, a research gap is identified and three research questions are derived, which lay the foundation for the thesis.

Chapter 3 introduces the theories consulted to derive the hypotheses for the empirical part of this study and to answer the formulated research questions. The theory of planned behavior builds the theoretical foundation of this study and is discussed in detail. A literature review about publications connecting the TPB and individuals' knowledge exchange behavior is presented and motivational factors that impact on employees' willingness to exchange knowledge in OI-projects are identified. Based on the TPB and the literature review, hypotheses are derived and a research model is composed.

In *chapter 4*, the research approach of this thesis is explained and the process of company selection for the empirical part of my study is outlined. In addition, details on the qualitative pre-study (interviews) and the quantitative main study (online survey) are provided. In particular, the development and pre-test of the questionnaire, the data collection procedures, and the data analysis methods are explained.

Chapter 5 summarizes the findings from interviews conducted with R&D managers and reveals their perspectives on open innovation. The typical procedure of setting up an OI-project and selecting an OI-partner is outlined and basic conditions for an OI-project are expounded. Furthermore, open innovation is assessed based on the identified advantages and disadvantages.

In *chapter 6*, the findings from the online survey are summarized. After discussing the data distribution and bias treatment, sample characteristics and some other interesting descriptive findings are highlighted. The results from an open-ended question regarding OI-requirements are then presented. Lastly, the measurement model and the structural model are evaluated.

Chapter 7 discusses findings from the interviews and online survey with regard to the three research questions. The findings are compared with and related to prior research to take a holistic view on the research questions and to answer them. Further, the literature is consulted to find possible explanatory approaches for hypotheses that were not supported by the data.

In *chapter 8*, the findings of my study are considered with regard to their contribution to academic research. Furthermore, managerial implications are derived. Lastly, the limitations of the study are highlighted and recommendations for further research are formulated.

2 Conceptual Foundation

This chapter outlines the underlying concepts of this study and research questions are framed, based on the identified research gap. The fundamental concept is the OI-approach. Therefore, the first sub-chapter provides information about antecedents and basic principles and gives an overview of prior and current research related to open innovation. Further, the link between open innovation and knowledge management is emphasized. As a result, OI-relevant aspects of knowledge management are discussed in the second sub-chapter. Finally, attention is drawn to the identified research gap and research questions for this study are compiled.

2.1 Open Innovation

The term open innovation can be traced back to the eponymous book of Chesbrough (2003), where he describes the opening up of the conventional, rather closed innovation process and introduces the OI-concept based on observations mainly from high-tech industries. He defines open innovation as '[...] *a paradigm that assumes that firms can and should use external ideas as well as internal ideas, and internal and external paths to market, as the firms look to advance their technology.*" (Chesbrough 2003, p. xxiv) He succeeded in creating a broadly known keyword for the integration of external sources into the innovation process, even though he was not the first to expound this idea.

This chapter provides insights into the OI-phenomenon, although it does not aim to deliver a comprehensive review. In the following, the shift from closed to open innovation is explained and basic principles of open innovation are presented. Thereafter, I give an overview of prior research related to the integration of external innovation sources. Lastly, developments in OI-research and different perspectives on open innovation are introduced.

2.1.1 From Closed Innovation to Open Innovation

The typical innovation process follows a stage-gate scheme (see Cooper 1990, 1996; Verworn and Herstatt 2000) and can be described as a funnel with a broad front end. The front end represents the research component of R&D, where ideas enter the process and start the invention. It follows the idea realization and development phase, where promising ideas are realized and the development part of R&D begins. In the commercialization stage, inventions are transformed into innovations and brought to the market. (cf. Herzog 2008, p. 11; Nelson and Winter 1982, p. 263; Schumpeter 1934, pp. 88f)

In the conventional, vertical integrated innovation process, all R&D efforts are centralized and take place solely in-house (see Chandler 1977). As shown in Figure 1, the company relies on its own knowledge base and only internal ideas gain access to the innovation process. The R&D department uses the most promising ideas to develop products, which are finally marketed and distributed by the company. Chesbrough calls this a "closed innovation" model, because ideas and products can enter and respectively leave the process only at one point. (cf. Chesbrough 2006c, p. 2)

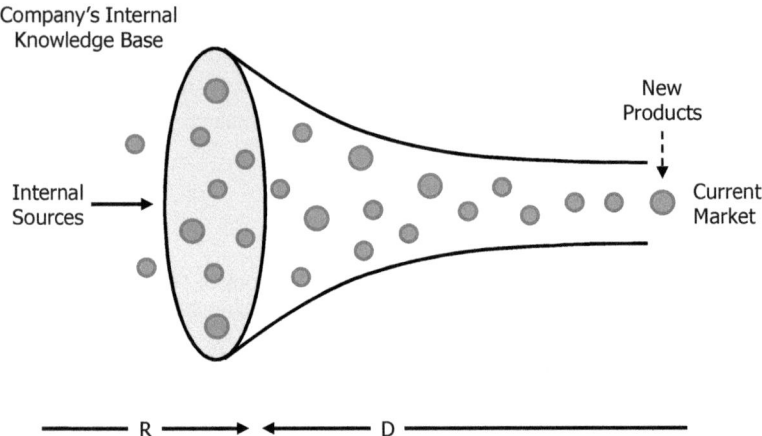

Figure 1: Closed Innovation Model[2]

For a long time, this approach proved very successful for companies. They heavily invested in their R&D and were able to achieve breakthrough innovations. They transformed them into new products and sold the products, which yielded higher margins and increased profits. In turn, these were reinvested in R&D (cf. Chesbrough 2003, p. xxi). This approach helped companies to grow, to protect and control their intellectual property resulting from the innovation process, and to enhance their knowledge base (cf. Chesbrough 2003, p. 34). However, the R&D centralization isolated the experts in one company from their peers in others. Over time, this encouraged R&D employees to believe high quality could only be achieved internally, which, in turn, promoted a preference for internal solutions (even if inferior to external alternatives) and created the NIH-syndrome[3] (cf. Chesbrough 2003, pp. 29f.; Möslein and Neyer 2009, p. 88). Clagett (1967) was the first scholar to address the NIH-syndrome directly. In his work, he described it as negative attitude of a technical organization towards ideas and innovations from outside the organization (cf. Clagett 1967,

---

[2] Illustration: cf. Chesbrough 2006c, p. 3
[3] The NIH-syndrome predominantly affects R&D employees and relates to external ideas/knowledge/ technologies coming into a company. With the Not-Sold-Here (NSH) syndrome, Chesbrough 2003, pp. 186ff. introduced the business counterpart to the R&D-typical NIH-syndrome. The NSH-syndrome is related to internal ideas/knowledge/technologies that leave the company to be used externally. It can best be described as a negative attitude of the business toward the outflow of internally developed ideas/ knowledge/technologies for external use.

p. ii). In a later article, the NIH-syndrome "[...] *is defined as the tendency of a project group of stable composition to believe it possesses a monopoly of knowledge of its field, which leads it to reject new ideas from outsiders to the likely detriment of its performance."* (Katz and Allen 1982, p. 7) This definition expands the first one, as it explains corporate resistance to adapting external ideas (belief in internal knowledge monopoly) and at the same time points out the consequences (impairment of company performance). The NIH-syndrome is likely to have a negative effect on company performance because the internal R&D depends on impulses coming from outside to keep pace with (technical) progress, i.e., R&D employees have to be able to gather and process information from external sources to increase the company's internal knowledge base and keep it up to date (cf. Katz and Allen 1982, p. 7). This implies it is impossible even for a company with the brightest employees to have all relevant knowledge and expertise in-house, which represents one of the constitutive assumptions of the OI-approach (cf. Chesbrough 2003, p. xxvi, 2006c, p. 2). Over the years, this became clear to more and more companies, which consequently began to open up their innovation processes. According to Chesbrough (2003, pp. 34ff.), this shift from closed innovation models to OI-models was facilitated in particular by the growing availability and mobility of experienced and qualified people. Advances in information and communication technologies and their increased availability further supported the establishment of the OI-approach in various industries (cf. Chesbrough and Brunswicker 2013, p. 6; Dodgson et al. 2006, pp. 333ff.; West and Bogers 2014).

Figure 2 depicts the OI-model.[4] The company border is permeable and allows interaction between the company and its environment during the entire innovation process. Contrary to closed innovation (see Figure 1), internal as well as external ideas can enter the innovation process at the front end. Furthermore, external impulses can also enter during later phases, e.g., the development stage (cf. Chesbrough 2006c, pp. 2f.). Another fundamental difference is that the OI-approach appreciates the outflow of promising ideas (e.g., in form of licensing or spin-offs), where ideas do not fit with the intended business model and would not be advanced in-house, but could flourish within other business models (cf. Chesbrough 2006c, p. 8). Therefore, Chesbrough (2006c, p. 1) extended his first definition of open innovation as follows: "[...] *Open Innovation is the use of purposive inflows and outflows of knowledge to accelerate internal innovation, and expand the markets for external use of innovation, respectively. Open Innovation is a paradigm that assumes that firms can and should use external ideas as well as internal ideas, and internal and external paths to market, as they look to advance their technology."* Furthermore, he underlined the central role of the business model (cf. Chesbrough 2006c, p. 8).

---

[4] The difference between inbound and outbound open innovation will be explained in chapter 2.1.3.1.

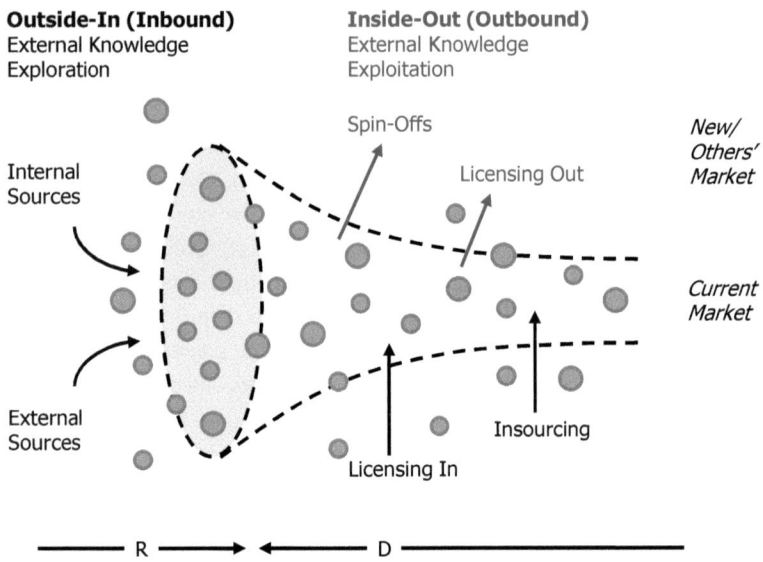

Figure 2: Open Innovation Model[5]

As already mentioned, the OI-concept assumes it is impossible for a company to have all of the required expertise and suitable knowledge in-house (cf. Chesbrough 2006c, p. 2). Useful knowledge is rather expected to be widely distributed and of generally high quality (cf. Chesbrough 2006c, p. 9). Moreover, internal and external knowledge is considered to be equally important for a company (cf. Chesbrough 2006c, p. 8).[6] Thus, knowledge exchange with external sources is necessary and valuable. In order to optimize the outcome of the innovation process, companies should try to find the appropriate balance between internal and external R&D (cf. Chesbrough 2003, p. xxvi).

2.1.2 Prior Research with Focus on External Innovation Sources

By introducing the OI-concept, Chesbrough succeeded in creating a broadly known term for the integration of external sources into companies' innovation processes and the appreciation of related knowledge inflows and outflows. Of course, Chesbrough's work was not detached from prior research and not all underlying ideas were completely novel. His research was based on – and is in line with – many previous studies. Due to this fact and the popularity of the open innovation term, it is not surprising some critics have described it as "[...] *old wine in new bottles.*" (Trott and Hartmann 2009, p. 1) But even if its actual degree

---

[5] Illustration: cf. Chesbrough 2006c, p. 3
[6] It is also considered equally important that innovations find internal as well as external ways (e.g., licensing, spin-offs) to market (cf. Chesbrough 2006c, p. 1).

of novelty is arguable, the OI-approach definitely complemented prior concepts by explicitly addressing some underlying principles, e.g., the equal importance of internal and external knowledge and the central role of the business model (see Chesbrough 2003). Furthermore, Chesbrough successfully labeled a collection of previous developments and research activities and so coined an umbrella term for a variety of phenomena:

"By giving it a label, it got a face, and the following stream of studies gave it a body too."
(Huizingh 2011, p. 3)

This made it easier especially for practitioners to communicate.[7] Although not everybody might have exactly the same understanding of open innovation, it at least established a common denominator. Due to scope of prior research, there now follows an overview of selected studies that contributed to the development of the OI-concept.

The importance for companies to establish external linkages was recognized relatively early. Hippel (1976) claimed users were an important source of innovation. In 1986, he introduced the lead user concept and so highlighted a specific group of users as a promising innovation source.[8] In his book "The Source of Innovation", published two years later, Hippel suggested that not only users but also other groups, such as suppliers, possess high innovative potential. In his eponymous article, Allen (1983) examined the phenomena of "collective invention" among firms. Teece (1986) demonstrated companies' need to connect with their business environment using the example of complementary assets. According to Cohen and Levinthal (1990), the ability to recognize valuable information from outside the company and to absorb and assimilate it is an essential capability (absorptive capacity), since external knowledge is crucial to companies' innovation process. That firms can benefit from permeable company boundaries, even if knowledge is exchanged with competitors, was argued by Schrader (1991). Jaffe (1989) and Lee (1996) examined the role of universities and academics in industrial innovation. Due to the increased interest in external technologies, Chatterji (1996) presented a code of practice for technology sourcing. In 1999, Raymond published an article about the open source concept and, thereby, triggered intensive research in that area.[9]

---

[7] However, some academics believe it has complicated academic communications (see Groen and Linton 2010; Hippel 2010; Linstone 2010).
[8] Interest in user-driven innovation, particularly the concept of lead users, rocketed in the following years, which resulted in countless publications (see e.g, Baldwin et al. 2006; Bogers et al. 2010; Franke et al. 2006; Herstatt and Hippel 1992; Lüthje and Herstatt 2004; Reichwald and Piller 2009; Schreier and Prügl 2008; Schweisfurth and Raasch 2012).
[9] Open source has its origins in the software industry and unlicensed handling of source code is central to this, i.e., developers are given the necessary access to advance, modify, and distribute the source code. In 1998, Raymond and a colleague established the Open Source Initiative (http://www.opensource.org/) to promote this idea. Further information on this initiative and a detailed definition of open source are provided on the website. For more information on related research see Feller and Fitzgerald 2002; Krogh and Hippel 2003; Krogh et al. 2003; Lakhani and Hippel 2003; Raasch et al. 2009.

Besides the importance of involving external sources during the process of innovation, scholars were also interested in the ways in which companies can connect with external partners and obtain external knowledge. As a result, two main research directions evolved: Strategic alliances[10] and in-sourcing, through mergers and acquisitions (see Hagedoorn and Duysters 2002). Strategic alliances in particular caught the attention of researchers (see e.g., Hagedoorn 1993; Lambe and Spekman 1997; Mowery et al. 1996; Narula and Hagedoorn 1999; Nicholls-Nixon and Woo 2003). According to Nooteboom (1999, pp. 64ff.), alliances can assume various forms. However, Mowery et al. (1996, p. 79) suggested two dimensions in order to classify them. Firstly, strategic alliances can be divided into equity-based (e.g., joint venture, minor equity investment) and contract-based (e.g., joint development agreement, R&D contract) collaborations. Secondly, strategic alliances can be categorized as unilateral (e.g., licensing), bilateral, or multilateral. Of all forms, joint ventures (i.e., equity based, bi- or multilateral alliance) were of special interest to researchers (see e.g., Harrigan 1986; Kogut 1988). However, Mowery et al. (1996, p. 80) showed this interest is not associated with how frequently joint ventures really occur in practice. In fact, the popularity of joint ventures decreased over time and contract-based alliances were gradually preferred (cf. Hagedoorn 2002, p. 478; Hagedoorn and Duysters 2002, p. 168). Independently from the mode of partnership, the motives for alliances were manifold and not always related to innovation and R&D (cf. Hagedoorn 2002, p. 477). For example, in the early 19th century alliances were used as vehicles for exploiting natural resources (cf. Mowery et al. 1996, p. 78). Later, companies aimed to establish technical standards and "dominant design", to shorten innovation cycles, to acquire new skills, to share the risks and costs of innovation, and to increase their market power (cf. Hagedoorn 2002, p. 480; Mowery et al. 1996, p. 79). Furthermore, globalization and other factors had increased the complexity of product development and the need for interdisciplinary expertise (cf. Narula and Hagedoorn 1999, p. 285). Regardless of the stated motive, Hamel et al. (1989, p. 134) noted that some alleged alliances can only be considered "[...] *sophisticated outsourcing arrangements."*

Many studies helped to lay the foundations for OI-research and assisted to "give it a face".[11] Nevertheless, it would be too narrow to assume open innovation is only the sum of these parts. Open innovation is far more than strategic alliances and, on the other hand, not all strategic alliances relate to innovation and can be considered open innovation. Open innovation is also related to the user innovation theory (see e.g., Hippel 1976, 1986, 1988), but not all aspects of this concept conform to the idea of open innovation (see West and

---

[10] In the literature, many synonyms for strategic alliances are used; among them strategic partnering, inter-firm alliance, collaboration, and co-operation.

[11] Prior studies from other research fields (e.g., spin-off decisions, mergers and acquisitions) might also have contributed to the development of the OI-concept. However, since this perspective is not the focus of this study, such prior studies are not discussed here.

Bogers 2010). This implies that prior concepts and the OI-model partially overlap, but are also partially exclusive (see Figure 3). Schweisfurth et al. (2011) revealed that the same is also true of different models within the OI-framework, e.g., open source innovation (cf. Hippel 2010, p. 555; Möslein and Bansemir 2011, p. 13).

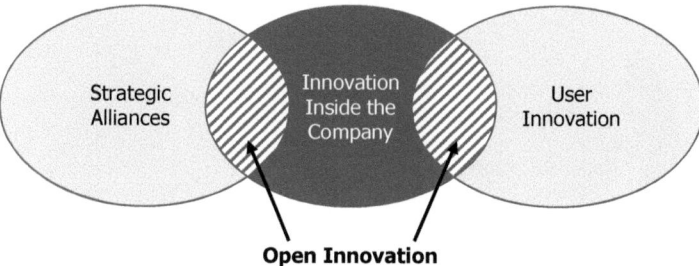

Open Innovation

Figure 3: Placement of Open Innovation Research[12]

Lastly, prior studies often considered internal and external sources as substitutes, which is inherent in the "make or buy" decision (cf. Tschirky et al. 2000, pp. 464f.) However, the OI-approach interprets internal and external knowledge as complementary and equally important (cf. Cassiman and Veugelers 2006, pp. 68ff.; Lichtenthaler 2011, p. 78; Reichwald and Piller 2009, p. 156; Schroll and Mild 2011, p. 490). Therefore, it accommodates the idea that internal R&D is essential for companies' ability to absorb and integrate external knowledge (see Cohen and Levinthal 1990; Veugelers 1997).

2.1.3 Current Developments in OI-Research

With his first book about open innovation, Chesbrough (2003) caught the attention of both practitioners and academics. His second book relating to open innovation (see Chesbrough 2006b) evoked a wave of OI-related studies dealing with different aspects of the concept. It became such a popular research topic that the R&D Management journal alone dedicated three special issues to open innovation.[13] This explosion of OI-related articles made it hard to keep track of all the developments within the field. Thus, numerous scholars tried to make a contribution to OI-research by reviewing and structuring the existing literature (see e.g., Dahlander and Gann 2010; Elmquist et al. 2009; Lichtenthaler 2011; Schroll and Mild 2012; West and Bogers 2014). These reviews revealed that quantitative research on open innovation is rare – especially in comparison to theoretical and qualitative studies – and

---

[12] Author's illustration
[13] R&D Management published special issues on open innovation in 2006, 2009, and 2010 (see Enkel et al. 2009; Gassmann 2006; Gassmann et al. 2010), but other journals (e.g., Industry and Innovation, 2008; International Journal of Technology Management, 2010; Research Policy, 2014; Technovation, 2011) also dedicated special issues to this topic (see Dahlander et al. 2008; Vrande et al. 2010; Huizingh 2011; West et al. 2014).

should take priority in future OI-research (cf. Dahlander and Gann 2010, p. 702; Lichtenthaler 2011, p. 80; Schroll and Mild 2012, pp. 86f.; Vrande et al. 2010, p. 225). Another finding was related to the level of OI-research. Generally, open innovation can be analyzed at different levels. Following Vanhaverbeke and Cloodt (2006, pp. 276ff.) and West et al. (2006, pp. 287ff.), six research levels can be distinguished (see Figure 4).

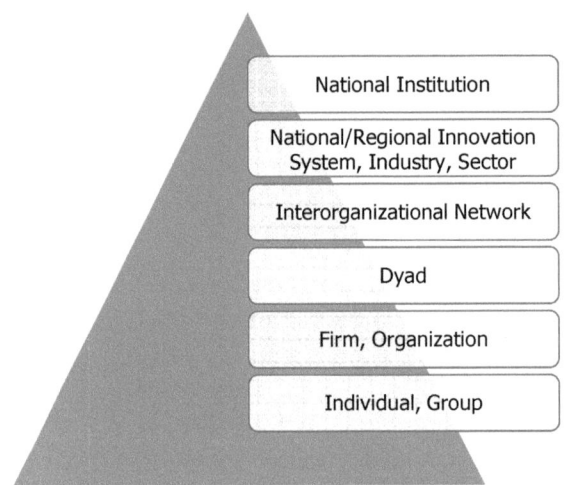

Figure 4: Levels of Analysis in Open Innovation Research[14]

The most elementary unit of analysis is the individual or group of individuals that make up a firm/organization (i.e., the second lowest level of analysis). OI-research can also focus on two companies connected through a strategic alliance (dyad level), for example. Furthermore, multiple interrelated parties can be the center of analysis (the inter-organizational network perspective). A fifth level of research relates to national/regional innovation systems, sectors, and industries. Lastly, OI-research can concentrate on national institutions.

Despite this range of possible OI-research levels, current studies have a clear emphasis and are not evenly spread across all layers. The majority of existing OI-studies concentrate on the level of firms/organizations[15] (cf. West et al. 2006, p. 287). Based on a set of 88 OI-related articles, Vrande et al. (2010, p. 226) found that more than 50% of the reviewed empirical studies focused on the firm level. The second largest share of studies (only 15%) dealt with innovation projects. OI-research related to the level of individuals is, however, rare and mainly focuses either on individuals engaged in open source projects and

---

[14] Illustration inspired by Vanhaverbeke and Cloodt 2006, pp. 276
[15] In the following, firm level, organizational level, and company level are all used interchangeably.

other OI-communities (for an overview see Schattke and Kehr 2009; see also the exemplary Fleming and Waguespack 2007; Hars and Ou 2002; Henkel 2009) or on lead users (see e.g., Franke et al. 2006; Lüthje 2004; Schreier and Prügl 2008). Very few studies address employee-related topics such as OI-relevant competencies and attributes (see Enkel 2010; Du Chatenier et al. 2010; Pedrosa et al. 2013) or possible barriers to open innovation such as the NIH-syndrome (cf. Enkel 2009, pp. 189ff.). In the study of Vrande et al. (2010, p. 226), only 11% of the articles under consideration were somehow related to the level of individuals. As a result of this imbalance, scholars tried to encourage other researchers to focus more on other levels of analysis and especially on individuals, since every (open) innovation starts with the effort of at least one individual (cf. Herzog 2008, pp. 3f.; Lichtenthaler 2011, p. 81; Vanhaverbeke 2006, pp. 206f.; Vanhaverbeke and Cloodt 2006, p. 279; Vrande et al. 2010, p. 230; West et al. 2006, pp. 287ff.).

The request for more research across all levels and especially on the individuals' level has also been prompted by norms regarding social theory building. According to Coleman (1990, pp. 2ff.), explanations on a macro-level should be based on examinations on the micro-level. Transferred to OI-research and the special interest in company-related issues, this claim echoes the demand for more research at the level of individuals. The argument is that if researchers are interested in open innovation at the firm level (macro-level), it is essential to understand the underlying reasoning and, thus, to involve the component parts of a firm in the research, as represented by its employees (micro-level). Figure 5 illustrates this argument and highlights the contention that it is inappropriate to directly draw conclusions from organizational antecedents on a given outcome. In fact, well-grounded theory has to have a micro-foundation, i.e., it starts and ends at the macro-level, but the arguments follow arrows 1, 2, and 3. The importance of the micro-foundation is further supported by other researchers (see e.g., Gavetti 2005; Teece 2007).

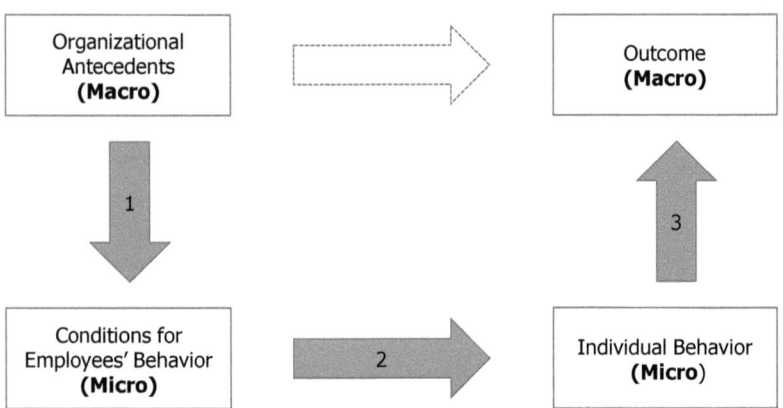

Figure 5: Macro- and Micro-Level Proposition[16]

In summary, increased research across different levels of analysis, especially at the level of the individual, is important for two reasons: Firstly, existing OI-studies have a strong focus on the organizational level and so lack consideration of other relevant layers. Secondly, the need for micro-foundation suggests it should start with the most elementary research unit: the individuals (i.e., employees of a company).

After shortly synopsizing existing OI-studies with respect to research type (i.e., theoretical, qualitative, quantitative) and level of analysis, the focus is on thematic aspects of current OI-studies. Consequently, I will now continue with the introduction of selected research streams in order to indicate the diversity of OI-related studies. This will be the foundation for the integration of my study into a broader research context. However, I do not aim to provide a comprehensive review of OI-research in general as it would extend the scope of this study.

2.1.3.1 Archetypes of Open Innovation Processes – Definition and Adoption

Based on Chesbrough's OI-conception, Gassmann and Enkel (2004) introduced three archetypes of OI-processes, which basically differ with respect to the direction of knowledge flows: outside-in (inbound), inside-out (outbound), and coupled processes.

In the case of inbound open innovation, knowledge flows from outside the company and enters its innovation process (see Figure 2). The underlying assumption is that the locus of knowledge creation can differ from the locus of innovation. Consequently, the outside-in process emphasizes the exploration of external knowledge or technologies. One way is to integrate external partners, e.g., in form of co-creations with customers (see Prahalad and Ramaswamy 2004), OI-alliances with competitors (see Han et al. 2012), and collaborations

---

[16] Illustration: cf. Coleman 1990, p. 8

with universities or suppliers. Another possibility is to source knowledge and technology through investments in external intellectual property (IP), e.g., in the form of acquisitions or in-licensing. Companies applying inbound-OI aim to enhance their internal knowledge base; improving their own innovation process; and increasing the success rate of products or projects. Gassmann and Enkel (2004, p. 10) suggested that inbound-OI takes place in low-tech industries in particular, where products are highly modular and knowledge intensity is high. (cf. Enkel et al. 2009, p. 312; Gassmann and Enkel 2004, pp. 7ff.; Gassmann and Enkel 2006, pp. 134f.)

In the case of outbound open innovation, knowledge flows from inside the company to the external environment, leaving the innovation process of the company (see Figure 2). The underlying assumption is that the locus of invention can be different from the locus of knowledge exploitation. Consequently, the inside-out process emphasizes the external exploitation of internal knowledge. Companies applying outbound-OI and transferring internal knowledge or technologies to the outside environment seek to bring products faster to the market than would be possible through internal development. This knowledge outflow enables companies to leverage their ideas and technologies and to realize additional profits by licensing or selling IP, transferring ideas/technologies to different applications and commercializing them in different industries (i.e., cross industry innovation). Outbound-OI is assumed to take place in large and/or mainly research-driven companies in particular. These aim to establish a technological standard, to get their products branded, or simply to decrease R&D-related fixed costs. (cf. Enkel et al. 2009, pp. 312f.; Gassmann and Enkel 2004, pp. 10ff.; Gassmann and Enkel 2006, pp. 135f.)

The coupled process basically combines the outside-in and inside-out processes, i.e., companies obtain external knowledge but also give internal knowledge to external players. The coupled process is often realized through co-operation with one or more complementary partners (e.g., strategic alliances, innovation networks). Reciprocity in terms of an appropriate, balanced, give-and-take relationship is crucial for this kind of OI-process. Therefore, all parties involved in a coupled process strive for a profound relationship with long-time interaction between the partners to facilitate intensive knowledge exchange and mutual learning. Companies applying the coupled process are assumed to aim for the establishment of a technological standard or dominant design. Thus, companies might be able to improve their competitive position and minimize risks. However, the coupled approach is not very likely to reduce development time, since close collaboration leads to increased co-ordination efforts. (cf. Enkel et al. 2009, p. 313; Gassmann and Enkel 2004, pp. 12f.; Gassmann and Enkel 2006, p. 136)

These three archetypes of OI-processes have laid the foundation for numerous studies. For example, Dahlander and Gann (2010) used the differentiation of inbound and outbound as the basis for their theoretical framework and developed it further by adding the dimension of pecuniary/non-pecuniary. As a result, they introduced four forms of openness: Acquiring (inbound, pecuniary); sourcing (inbound, non-pecuniary); selling (outbound, pecuniary); and revealing (outbound, non-pecuniary). Beyond these definitions they derived advantages and disadvantages for all four types based on a literature review. Another example is Chesbrough and Crowther (2006), who also based their analysis on the concept of inbound and outbound. They conducted interviews to determine if companies outside high-technology industry had adopted open innovation and which OI-activities were primary used. They found open innovation had been adopted and that the surveyed companies had a clear focus on inbound activities.[17] Furthermore, the study suggested reasons for adopting inbound-OI and obstacles related to the adoption. By applying a survey, Lichtenthaler (2008) took the research of Chesbrough and Crowther one step further and used the inbound and outbound dimension to assess companies' OI-strategies. He measured the extent to which companies acquire external technologies (inbound) and externally exploit their technologies (outbound) to conduct a cluster analysis. As a result, six clusters were identified, which basically represented four different strategies or innovator types: Companies were either closed innovators (clusters 1 and 2); open innovators with inbound emphasis (cluster 3); open innovators with outbound emphasis (cluster 4); or focused on coupled processes (clusters 5 and 6).[18] Comparing the number of companies focusing on inbound, outbound, and coupled processes respectively, the coupled OI-approach appeared to be the most frequently used strategy, followed by inbound, with outbound processes coming last (cf. Lichtenthaler 2008, pp. 150ff.). Based on the same data set, a second article was published, where technology aggressiveness was related to inbound and outbound activities (see Lichtenthaler and Ernst 2009). Furthermore, a positive relationship between inbound and outbound open innovation was hypothesized and supported by the data, i.e., the data suggested that an increase in inbound activities positively influenced outbound activities and consequently lead a company to adopt the coupled OI-process. This is in line with the result from Lichtenthaler (2008) and Schroll and Mild (2011), which found most companies apply coupled-OI and that pure outbound-OI is very seldom applied. However, the findings also suggest that companies' clear favoring of the integration of external knowledge means inbound and outbound activities are not balanced, even if the companies follow the coupled approach (cf. Schroll and Mild 2011, p. 489). A last study I would like to mention focused on open innovation in small and medium-sized enterprises (SMEs). Vrande et al. (2009) examined the adoption of open innovation in SMEs and found manufacturers as well as service providers increasingly open up their innovation processes and have a preference for inbound-OI, even though significant differences between small and medium-sized companies were identified.

---

[17] In the literature, inbound processes are looked at more frequently than outbound ones (for an overview cf. Schroll and Mild 2012, pp. 101ff.). Schroll and Mild 2012 generally emphasized that OI-adoption was of immense interest and an important topic in quantitative OI-research.
[18] The vast majority (67.5% of companies) belong to the cluster of closed innovators.

Furthermore, SMEs do not apply the whole range of OI-activities (e.g., only a few surveyed SMEs used in- and out-licensing or venturing activities).

The selection of studies discussed above demonstrates that open innovation is adopted across different industries, countries, and company sizes and that all three archetypes of OI-processes can be found in praxis. Furthermore, the studies indicate a preference for the coupled process, followed by pure inbound-OI. However, especially the article of Lichtenthaler (2008) does not only provide information about the primarily used OI-practices, but also point to the extent of OI-adoption as a relevant topic. Therefore, the next chapter will focus on some related studies and the degree of openness.

2.1.3.2 Degree of Openness

The degree of openness can either be interpreted as the extent of inbound and outbound OI-activities or as the extent of knowledge and IP control.[19] Researchers have shown interest in both interpretations. Therefore, I will discuss studies related to both forms of openness, starting with articles that focus on the extent of OI-activities.

With reference to the work of Katila and Ahuja (2002), Laursen and Salter (2006) introduced one of the most prominent scales for measuring the degree of openness. They suggested using two dimensions to assess a company's degree of openness: external search breadth and external search depth. The external search breadth is "[...] *defined as the number of external sources or search channels that firms rely upon in their innovative activities.*" (Laursen and Salter 2006, p. 134) The search depth indicates how intensively companies draw from their different external partners providing innovative ideas. Thereby, companies can choose from multiple knowledge sources (cf. Albers et al. 2000, pp. 95f.; Fey and Birkinshaw 2005, p. 601; Hippel 1988, pp. 28ff.; Neyer et al. 2009, p. 411). Universities and research institutes (see Tether and Tajar 2008; Laursen and Salter 2004), users and customers (see Herstatt and Lüthje 2005; Lettl et al. 2006; Reichwald and Piller 2009), suppliers (see Li and Vanhaverbeke 2009; Remneland-Wikhamn et al. 2011), competitors (see Lim et al. 2010), and partners such as non-profit organizations (see Holmes and Smart 2009) are regularly considered potential OI-partners.[20] Some scholars generally assumed that a greater diversity of partners would positively influence companies' innovation performance (see Faems et al. 2005; Nieto and Santamaría 2007). However, Laursen and Salter (2006) examined the relationship between the degree of openness and company's innovation performance and assumed curvilinear relationships between search breadth and

---

[19] If the degree of openness is gauged by the extent of inbound and outbound OI-activities, it can be linked to the concept of OI-maturity (see Enkel et al. 2011; Habicht et al. 2012).
[20] Some scholars also consider employees from business units other than R&D as potential OI-partners (cf. Chesbrough and Brunswicker 2013, p. 15; Möslein and Neyer 2009, pp. 89ff.; Schweisfurth 2013, p. 4). However, this study focuses on OI-partners outwith the company structure, i.e., OI-partners in this study refer to external partners.

innovative performance as well as between search depth and innovative performance. Both hypotheses were supported by the data, implying that on the one hand openness positively influences a company's innovative performance to a certain degree, but on the other hand the positive effect of openness has its limits. Following the results of Laursen and Salter, strict in-house innovation is not a dominant strategy but nor are too many external partners and/or too many close relationships with different external sources preferable for innovative performance. Consequently, companies walk a thin line and have to figure out their optimal degree of openness.

The idea of measuring the degree of openness based on search breadth and search depth has inspired a lot of other scholars who adopted the scale and partially adapted it for their studies (see e.g., Chiang and Hung 2010; Chen et al. 2011; Keupp and Gassmann 2009; Lee et al. 2010). Nevertheless, other approaches to measure openness have also evolved, particularly as a result of Laursen and Salter's focus on inbound-OI. Barge-Gil (2010), for example, assessed the degree of open innovation based on two criteria. Firstly, companies had to provide information on how innovations were usually developed (primarily alone, by collaboration, or from third parties). Secondly, companies had to assess the importance of 11 different knowledge sources (the internal knowledge base plus ten external partners). Depending on how companies had answered these two questions, they were classified as open, semi-open, or closed innovators. Another different approach was taken by Vrande et al. (2009). They did not measure an overall degree of openness, preferring to individually assess eight OI-practices (e.g., venturing, customer involvement, in- and out-licensing).

As demonstrated, degrees of openness can be measured in many ways. This complicates the aggregation and comparison of different findings related to, firstly, the influence of openness on aspects such as innovative performance and, secondly, the effect of factors such as company size and industry on openness. Despite these difficulties, Drechsler and Natter (2012) tried to better understand the reasoning behind companies' openness decisions. They hypothesized innovation strategy (i.e., internal R&D and knowledge acquisition), scarce company resources (i.e., financial and knowledge gaps), the appropriability regime (i.e., formal and strategic IP protection), and market dynamics (i.e., technology change, demand uncertainty, and increased competition) to influence a company's degree of openness in a predominantly positive manner.[21] Only increased competition was expected to negatively influence the degree of openness. If companies followed a closed innovation

---

[21] The degree of openness was measured by asking companies to rate the importance of knowledge from universities, public research institutes, private research institutes, customers, suppliers, and competitors and summing up the ratings (cf. Drechsler and Natter 2012, p. 442).

approach, Drechsler and Natter found most relations supported by the data.[22] However, the results were different for companies that already exhibited a certain degree of openness. Here, the market dynamics did not show any effect on openness. Rather, a higher level of openness related to a company's innovation strategy (knowledge acquisition was found to be complementary to internal R&D), financial gaps, and strategic IP protection.

The understanding of openness as discussed above is related to the intensity of inbound and outbound activities, i.e., the degree of openness depends on the number of external partners and the extent to which they are integrated into a company's innovation process (inbound); and/or how many and to what extent external paths to market are adopted by a company (outbound). Another interpretation of openness is related to the extent of knowledge control, i.e., whether innovators aim for IP protection (via patent or copyright) or freely reveal their knowledge. IP protection and free revealing are, thereby, based on different theories and represent the two extremes of a continuum, where IP protection stands for limited openness and free revealing for total openness.

The need to protect IP arises from the private investment model, which dictates that individuals or organizations will only invest in the development of an innovation if they can expect a greater pay-off than possible free riders (cf. Hippel and Krogh 2003, pp. 213f., 2006, p. 304). To encourage such private investment, society allows innovators to protect their knowledge and benefit from resulting patents and copyrights through licensing or selling them (cf. Hippel and Krogh 2006, p. 302). This applies in particular to profit-seeking individuals/organizations. It is an important element of the OI-approach according to Chesbrough, where efficient IP management plays a central role and in-licensing and out-licensing are considered as valuable options for obtaining external knowledge and generating additional revenues, respectively (see Chesbrough 2003, 2006c).

The origins of the concept of free revealing, on the other hand, can be found in the collective action model, which assumes that market failures induce innovators to release control of their knowledge, to share their ideas, and to convert them to public goods (cf. Hippel and Krogh 2003, p. 215; Hippel 2006, p. 89). Free revealing is therefore closely related to open science (see Dasgupta and David 1994), open source innovation[23] (see Raymond 1999; Schweisfurth et al. 2011), and Hippel's interpretation of openness, where innovators voluntarily giving up their IP rights so all parties have equal access and the innovation becomes a public good (see Harhoff et al. 2003; Hippel 2006; Hippel and Krogh 2006; Hippel 2010). However, this does not imply the acquisition and use of such knowledge

---

[22] Unexpectedly, companies' knowledge gaps negatively influenced the degree of openness. This was ascribed to a lack of absorptive capacity (see Cohen and Levinthal 1990). Demand uncertainty did not show any effect on companies' openness.

[23] Motives for freely revealing knowledge are most deeply examined in the context of open source software projects (cf. Hippel and Krogh 2006, p. 301). Furthermore, the question regarding the appropriate degree of openness was raised in particular in the contexts of open source platforms and communities (see e.g., West 2003). However, I will not go into further detail, since open source software is not the focus of this study.

has to be free of cost for potential adopters, nor that the benefits of applying the knowledge always exceed the acquisition costs. As long as the innovator or knowledge provider does not benefit from the expenditures made by the knowledge adopters, the knowledge can be considered as freely revealed (cf. Hippel and Krogh 2006, p. 296). This suggests that free revealing is not an alternative for profit-seeking companies. Indeed, the concept of free revealing was first mentioned in connection with users. However, it has also been noted among profit-seeking and even competing companies (see e.g., Allen 1983; Henkel 2003; Nuvolari 2004). Consequently, there have to be good reasons for free revealing. Hippel highlighted three practical cases, where free revealing is feasible and possibly superior to IP protection (cf. Hippel 2006, pp. 80ff.; Hippel and Krogh 2006, pp. 297ff.) Firstly, innovators cannot always autonomously decide between revealing and not revealing their idea, because other people might know something close to this idea, which could be used as a substitute. Therefore, the choice is often made between revealing voluntarily or involuntarily, since swift imitation is probably inevitable. Secondly, the ability to benefit from IP protection through patenting is limited. On the one hand, patents are not always worth the effort (time and money), on the other hand, not everything worth protecting can be protected via patenting (e.g., patents are rarely granted for ideas or mathematical formulas).[24] Thirdly, Hippel noted that innovators can directly benefit from free revealing in multiple ways. Free revealing of ideas and knowledge can increase the innovator's reputation, accelerate collective learning processes and, thus, push development. Furthermore, increased diffusion of innovation facilitates networking effects[25] and enables the establishment of dominant designs and/or technical standards, which leverage the commercialization of later versions. However, one of the disadvantages of this innovation model is that free riders benefit almost equally, which negatively affects innovators' motivation and so complicates the recruitment of contributors (cf. Hippel and Krogh 2006, p. 302).[26]

Using the example of open source software, Hippel and Krogh (2003) demonstrated that innovations are not necessarily the result of an either-or-decision between the private investment model (i.e., IP protection) and the collective action model (i.e., free revealing), but rather possess public as well as private elements. Based on this observation, they introduced the private-collective model, which combines the best aspects of both previously discussed innovation models, i.e., innovators invest in the development of public goods because they can expect higher benefits than free riders. The private-collective model, thus,

---

[24] In some fields (e.g., chemicals, pharmaceuticals), patenting is generally considered useful (see Arora et al. 2001).
[25] The networking effect basically means the value of a product or service increases the more people use it. A typical example would be telephony – the more handsets in circulation, the more valuable each telephone in use (cf. Hippel and Krogh 2006, p. 301).
[26] Using the example of open source software, West and Gallagher 2006 identified three main challenges relating to open innovation: Maximizing returns to internal R&D; incorporating external knowledge; and motivating individuals/organizations to generate and contribute their knowledge.

represents a middle course between the private investment and the collective action model. Henkel (2006) named this mix: "selective revealing".

In summary, the degree of openness can be viewed from at least two perspectives. On the one hand, it can provide information about the extent to which a company has adopted inbound and/or outbound OI-activities. In this case, the smallest degree of openness is somewhat greater than closed innovation and the highest degree is represented by companies that extensively apply different inbound as well as outbound OI-activities (but also invest in internal R&D). On the other hand, the degree of openness can be related to different forms of IP control. Here, the smallest degree of openness occurs when innovators protect their IP at its best and profit from patents through licensing. The highest degree is represented by free revealing of knowledge and has no IP protection at all.

2.1.3.3 Open Innovation Lifecycle and Essential Capabilities

After discussing different archetypes of OI-processes and degrees of openness, I will continue with the lifecycle of open innovation. In this context, I will introduce four selected studies dealing with the different phases of mainly inbound-OI and required capabilities. This sub-chapter aims to illuminate the phases companies with inbound-OI typically have to pass. Furthermore, it seeks to shed light on the capabilities necessary for (inbound) open innovation.

The first article I would like to highlight was written by Zahra and George (2002). The study was published before the introduction of the OI-term in 2003. Therefore, it does not directly address open innovation, but the closely related concept of absorptive capacity. Zahra and George defined four capabilities or dimensions that comprise absorptive capacity and can be interpreted as steps of inbound-OI: acquisition, assimilation, transformation, and exploitation. During the acquisition phase, companies identify and acquire relevant knowledge from external parties. In the second phase, companies analyze the acquired information and try to process, interpret and understand it. During the transformation phase, companies combine the newly obtained and assimilated information with existing, internal knowledge. Lastly, companies exploit their leveraged or newly created competencies. (cf. Zahra and George 2002, pp. 189f.)

Based on his experience from praxis and research for other practitioners in particular, Slowinski (2005) introduced the "want, find, get, manage framework". During the "want" phase, companies define the knowledge and resources needed for growth and evaluate the trade-offs between internal R&D and external sourcing. The "find" phase is concerned with localizing appropriate external sources that could provide the needed knowledge. The result of this phase is a short-list of potential partners, representing the basis for the "get" phase. In this third phase, companies evaluate the best potential source from the short-list for an OI-partnership. This decision is based on the expected knowledge gain from the partnership, the degree of common interests, and the prospects of mutual agreement. However, it is also possible the short-list will comprise only one potential partner (e.g., the expert in the field of interest), which naturally abbreviates the "get" phase. The last step follows if contracts and mutual agreements are signed and the partnership is officially established. Then the actual work begins. The resources and competencies of all partners have to be coordinated and integrated to accomplish the goals of the OI-partnership. Special attention has to be paid to a common understanding, i.e., all employees involved in the OI-project should be clear about who is responsible for what and the precise knowledge to be exchanged. (cf. Slowinski and Sagal 2010, pp. 39ff.)

Lichtenthaler and Lichtenthaler (2009) brought together open innovation and knowledge management by considering corporate internal and external knowledge exploration, knowledge retention, and knowledge exploitation. Based on these two dimensions (i.e., internal/external and exploration/retention/exploitation), they formulated six capacities required during the OI-process. The internal exploration of knowledge takes place in companies' R&D departments, where employees generate new knowledge. Therefore, inventive capacity is needed in this phase. External knowledge exploration means companies acquire necessary expertise from external partners, which requires absorptive capacity. Internal knowledge retention focuses on the internal storage and retrieval of knowledge, whereas external knowledge retention refers to the maintenance of knowledge in inter-organizational partnerships (e.g., alliances). For the internal retention, transformative capacities are needed. The external retention requires connective capacities. If knowledge is internally exploited, it is used for the development of a company's own products, requiring innovative capacity. Where there is external knowledge exploitation, knowledge leaves the company by means of licensing or spin-offs and desorptive capacity is required. Companies that apply the coupled process are engaged in all six phases and require all of the listed knowledge capacities. Since phases and related tasks are not completely sequential, companies must also have the ability to dynamically manage their knowledge base and processes over time.

The last article I would like to mention was written by West and Bogers (2014). In order to systematically review a selection of OI-related studies, they developed an integrative, four-phase OI-model. The first phase is concerned with obtaining knowledge from external sources. More precisely, companies directly search for potential external partners or facilitate the search (i.e., they turn to technology scouts, intermediaries, online communities, etc.; see chapter 2.1.3.4), filter the most promising partners, and finally acquire the relevant knowledge by signing licensing agreements or other contracts. In the second phase, companies integrate the newly obtained knowledge into their R&D activities. In this phase, company culture plays a central role, especially if NIH-tendencies have to be overcome (see chapter 2.1.3.5). Furthermore, absorptive capacity is essential at this point. After companies have integrated the external knowledge into their R&D process, they try to commercialize the resulting innovations. One of the main challenges in this phase is to align the selection of innovation and the corresponding commercialization strategy with the company business model. The first three phases of West and Bogers' model can be related at least partially to phases of the other three models suggested by Zahra and Georges (2002), Slowinski (2005), and Lichtenthaler and Lichtenthaler (2009). The fourth phase, however, is special because it is completely dedicated to the interaction mechanisms between OI-partners during an OI-process. These interaction mechanisms are iterative processes and, therefore, disrupt the otherwise linear model.[27] Examples are feedback loops (see e.g., Berkhout et al. 2006; Mortara et al. 2010) and reciprocal knowledge exchange in alliances, networks, and communities.

Comparing the four models suggested by Zahra and George, Slowinski, Lichtenthaler and Lichtenthaler, and West and Bogers, it becomes clear the authors all have a similar understanding of the OI-lifecycle. However, none of the models is completely congruent with any other. They vary with respect to number, granularity, and phase names. As indicated in Figure 6, a phase of one OI-lifecycle model is sometimes only one segment of another model's phase and vice versa. However, the level of aggregation is generally high, so that each phase in any of the models could be subdivided into multiple smaller steps. This already implies the difficulty of proposing one true and always appropriate OI-lifecycle model. Nevertheless, the synthesis of the four models results in a comprehensive OI-lifecycle model with five phases.

According to this model, companies with inbound-OI firstly define the knowledge or expertise needed to achieve a certain goal; deliberate whether they possess it internally; and evaluate the trade-off between internal R&D and external sourcing. The realization external knowledge is needed is the kick-off point for the second phase. The obtaining phase starts with the search for appropriate partners and ends with signing contracts – or at least agreements – with the partner(s) that fits best. After all parties have signed, the actual knowledge flow starts and this has to be properly managed. The third phase is concerned with integrating the newly obtained knowledge into the company, i.e., companies analyze the novel information and try to make sense out of it so the externally acquired knowledge can be merged with the internal knowledge base. This is the pre-condition not only for exploiting the knowledge, but also for storing and retrieving it later on. The combination of external and internal knowledge results in an innovation that can be exploited in the fourth phase. The most common way to exploit this innovation is to bring it to market and generate profits by selling it. The last phase is non-linear – related to interaction mechanisms between

---

[27] The idea that OI-processes do not necessarily follow a linear structure is not new (see e.g., Lichtenthaler and Lichtenthaler 2009), but designating a phase to these iterative processes represents a peculiar approach.

the OI-partners – and can occur during the entire OI-lifecycle. Since each OI-project requires a certain degree of interaction between the partners, sooner or later this phase becomes part of each OI-process.

In summary, the sub-chapter indicates the complexity of (inbound) open innovation and shows a properly executed OI-approach requires a lot of care and capabilities.

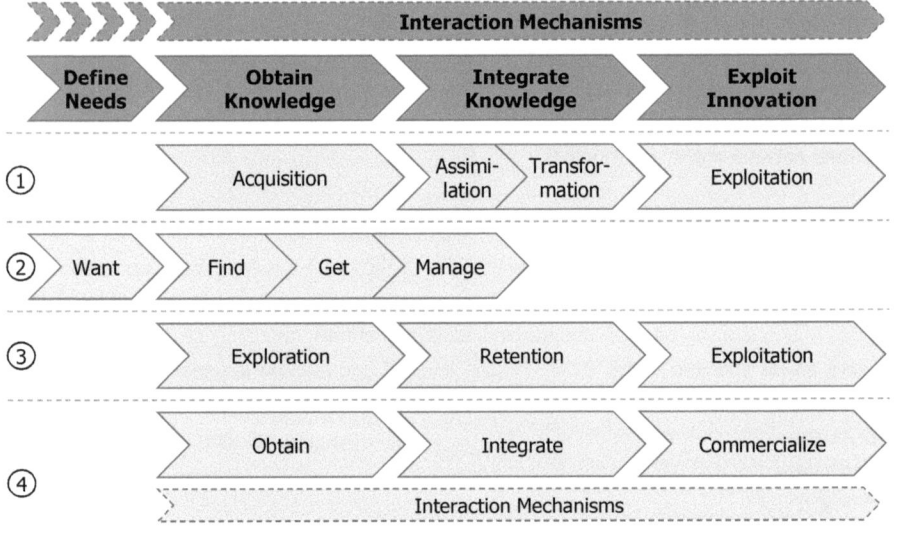

① Zahra, George (2002)　② Slowinski (2005)　③ Lichtenthaler, Lichtenthaler (2009)　④ West, Bogers (2013)

Figure 6: Phases of (Inbound) Open Innovation[28]

2.1.3.4 Open Innovation – Tools, Techniques, and Related Concepts

Many researchers and practitioners have shown particular interest in the various existing tools, techniques, and concepts related to open innovation. Since these methods are not the focus of my study, I will not go into great detail. However, it is important to give a short overview of the most prominent ones (which are mostly employed during the obtaining phase), because these tools, techniques, and concepts epitomize open innovation for many people.

A concept often directly linked to open innovation is crowdsourcing (see Schweisfurth et al. 2011). The term was coined in 2006 by Howe (2006b), even though the idea was not new at the time (cf. Afuah and Tucci 2012, p. 355). Crowdsourcing stands for the outsourcing of tasks to a mostly anonymous crowd. Therefore, the tasks are not directly assigned to

---

[28]　Author's illustration

individuals. It is an open call to participate in the problem-solving process. Individuals self-select to work on the task and so to be part of the crowd (see Afuah and Tucci 2012; Howe 2006a; Pirker et al. 2010). Consequently, crowdsourcing is a great opportunity to use the intelligence of many motivated people. However, not every problem is suitable for crowdsourcing. It has to be modular to a certain degree and possible to delineate and transmit it (see Afuah and Tucci 2012). The tasks do not necessarily need to be related to innovation and the encouraged crowd is not always external to the organization (see Pirker et al. 2010), so crowdsourcing cannot simply be equated with open innovation. However, both concepts overlap to a significant degree (similar to strategic alliances and user innovation, see Figure 3).[29]

To tap into crowdsourcing, companies can choose different approaches (see Boudreau and Lakhani 2013; Pirker et al. 2010). One very frequently applied technique is an ideas contest or ideas competition (see Boudreau and Lakhani 2013; Habicht et al. 2011; Möslein and Neyer 2009; Piller and Walcher 2006; Walcher 2009). The company releases a task, sets a deadline and the crowd can submit their solutions. When the contest is finished all of the proposals are evaluated by a group of experts and the person with the best solution wins the award. Companies can conduct such contests on their own or they can elicit assistance. Thereby, companies could receive support from many different knowledge brokers and for different problems (see Lichtenthaler and Ernst 2008; Sieg et al. 2010; Verona et al. 2006). With respect to open innovation and crowdsourcing, companies often refer to innovation intermediaries or innovation marketplaces such as InnoCentive or NineSigma[30] (see Burkhart et al. 2010; Habicht et al. 2011; Lakhani 2008; Möslein and Neyer 2009; Sieg et al. 2010). These intermediaries help to formulate the problem and provide a crowdsourcing platform where companies can anonymously post their challenges to a pool of talented and motivated people and only have to pay for solutions that fit the predefined criteria. Another way of using crowd intelligence is to establish and involve an online community (see Boudreau and Lakhani 2013). Schweisfurth et al. (2011) argue this technique has a certain overlap with the concept of crowdsourcing but is more closely related to the concept of open source innovation, which they consider a distinct research stream. A final way to integrate the crowd into a co-creation or co-design process and get access to "sticky information"[31] is through the application of toolkits (see Hippel 2001; Hippel and Katz 2002; Jeppesen 2005; Reichwald and Piller 2009). A toolkit targeting users has to fulfill five criteria: It has to be user friendly, meaning that no special skills or additional training is required to use the toolkit competently. A library of frequently used modules should also be part of the toolkit so

---

[29] For further reading on crowdsourcing, refer to Boudreau and Lakhani 2013; Howe 2009; Jeppesen and Lakhani 2010; Poetz and Schreier 2012.
[30] For InnoCentive see: http://www.innocentive.com/ and for NineSigma see: http://www.ninesigma.com/ (both last accessed on May 30, 2013)
[31] According to Hippel 1994, information is "sticky" if its acquisition, transfer, and use are costly. For a company, information about customers' needs is often sticky.

that users can easily incorporate these modules into the product. Thirdly, the toolkit has to limit users' creativity to a certain solution space, without excluding important aspects they might want to integrate into the product. Fourthly, users should have the chance for trial-and-error learning during the whole process of their product design and development. Lastly, the toolkit has to be designed in such a way that all products deriving from it can be produced without the manufacturer having to adapt existing production equipment. (see Hippel 2001; Hippel and Katz 2002)

Toolkits are also frequently applied in the process of mass customization (see Piller et al. 2004; Pine 1993; Reichwald and Piller 2009), which is sometimes confused with open innovation or crowdsourcing. Although mass customization also uses a crowd (customers), people are not involved in the innovation process. Rather, they are integrated into the configuration and design phase, i.e., they are invited to customize their own product (cf. Reichwald and Piller 2009, p. 53). That way, many companies doing business in the consumer goods' market hand over design-related tasks to their customers. Famous examples of companies with such mass customization tools on their websites include Adidas and Lego.[32]

In summary, companies can choose between different approaches to integrate the crowd and their customers in particular. However, some business models do not opt for one approach, preferring to embrace several techniques. A good example is the fashion company Threadless[33], where customers become co-creators and undertake the design of shirts the company is selling. The company has almost completely outsourced this part of the value chain. People are invited to design the shirts using a toolkit and to submit their designs. After submission, other customers evaluate the proposals and commit to buying a shirt with this design. Threadless will only commit to producing the design if a certain number of individuals have committed to buying it. Moreover, they will only make as many as ordered.[34] As soon as a shirt goes into production the customer who designed it will receive a monetary reward and all shirts with this design will have a label with his/her name printed on it. (cf. Piller 2010, pp. 2ff.; Reichwald and Piller 2009, pp. 2f. and company website)

2.1.3.5 Barriers to Open Innovation

Companies have good reasons to engage in open innovation – the advantages are obvious and frequently highlighted (see e.g., Chiang and Hung 2010; Dahlander and Gann 2010; Ili and Albers 2010; Laursen and Salter 2006; Reichwald and Piller 2009; Vanhaverbeke et al. 2008). However, open innovation also has disadvantages that might discourage companies from embracing this approach (see e.g., Knudsen and Mortensen 2011; Lokshin et al. 2011).

---

[32] For Adidas see http://www.miadidas.com/ and for Lego see: http://www.lego.com/en-us/createandshare/ (both last accessed on May 30, 2013)
[33] See http://www.threadless.com/ (last accessed on May 30, 2013)
[34] Reproduction of the design is only possible when a critical mass of customers pledge to buy it.

Barriers could exist, particularly at the level of individuals and among employees. Behrends (2001, p. 96) and Hauschildt and Salomo (2011, pp. 125f.) distinguished three types of barriers: Individuals do not want ("want-barrier"), are not able ("can-barrier"), and/or are not allowed ("shall-barrier"). Applied to open innovation, the "want-barrier" relates to motivational aspects and refers to employees' willingness to be involved in OI-activities. The "can-barrier" describes the lack of necessary capabilities and know-how to cope with the challenge of open innovation. The "shall-barrier" refers to employees' perception that open innovation is not desired by other key persons or groups (cf. Enkel 2009, pp. 189f.; Haller 2003, p. 192).[35] Enkel (2009, pp. 189ff.) made an effort to gather reasons responsible for the development of such barriers in the context of open innovation:

A central point of the "want-barrier" and a major obstacle for open innovation is the existence of the NIH-syndrome (cf. Elmquist et al. 2009, p. 337; Herzog 2008, p. 197; Lichtenthaler and Ernst 2006, p. 369; Piller and Reichwald 2009, pp. 117f.; Wecht 2006, pp. 174f.).[36] Furthermore, employees might be put off by the bureaucratic and administrative efforts connected with OI-projects. Employees may also believe there is risk they might "lose" their knowledge and, therefore, their expert status. The prospect of sharing the IP with an external partner might also add to the development of a "want-barrier". A last point highlighted by Enkel (2009) is related to the willingness of external partners. Not only the individuals within the company have to be motivated to participate in open innovation, but also the external individuals (e.g., customers, suppliers). (cf. Enkel 2009, pp. 189ff.; Riege 2005, pp. 23ff.; Vrande et al. 2009, pp. 433ff.)

The "can-barrier" is fostered by the lack of technical skills and other capabilities. For example, OI-projects place special demands on employees' administrative, communicative, organizational, leadership, and boundary spanning capabilities. Furthermore, the "can-barrier" is also closely related to missing absorptive capacity and open-mindedness. (cf. Enkel 2009, pp. 189ff.; Habicht et al. 2011, pp. 61f; Riege 2005, pp. 23ff.; Vrande et al. 2009, pp. 433ff.)

Finally, the "shall-barrier" develops if management is not committed to open innovation and fails to support OI-activities adequately. This is often associated with an imbalance between daily business and OI-activities. Employees engaged in OI-projects often need some degree of freedom with respect to time and money. If the company is not committed to open innovation, it is easy to build a barrier by restricting the financial resources and time

---

[35] Gemünden and Walter (1996, pp. 237f.) and Schüppel (1999, p. 35) have added the "know-barrier" – a fourth obstacle entailing the lack of awareness on the part of an individual of the appropriate contact person within the OI-partner's organization or where an individual isn't sure how to contribute to the OI-project. However, this barrier is not discussed further in this thesis, as it is not completely distinct from the "can-barrier".

[36] Lichtenthaler and Ernst 2006 extended the NIH-syndrome; suggesting five more syndromes that should be taken into consideration.

available for OI-projects. (cf. Enkel 2009, pp. 189ff.; Riege 2005, pp. 23ff.; Vrande et al. 2009, pp. 433ff.)

Enkel (2009) focused on barriers in the context of open innovation, i.e., where a company opens its boundaries and exchanges knowledge with at least one external partner. However, Grote (2010, pp. 33ff.) pointed in his dissertation to the existence of similar barriers when it comes to co-operation and knowledge flows among independent business units within a company. He empirically demonstrated the importance of the encouragement of such co-operation through incentive systems and integration mechanisms. It can be assumed though that the impact of such facilitating factors might differ in inter- and intra-organizational relationships (see Li 2005).

2.2 Knowledge Management

The knowledge inflows and outflows during an innovation process are central to Chesbrough's definition of open innovation (cf. Chesbrough 2006c, p. 8), indicating that open innovation is associated with the management of knowledge and especially with knowledge exchange. Lichtenthaler and Lichtenthaler (2009) reinforced this link. In a later article, Lichtenthaler (2011, p. 77) defined open innovation *"[...] as systematically performing knowledge exploration, retention, and exploitation inside and outside an organization's boundaries throughout the innovation process."*[37] Consequently, knowledge management is closely related to open innovation and an essential task for companies that adopt the OI-approach.

In this chapter, I will give an overview of basic knowledge processes regularly discussed in the knowledge literature. Following this, I will further detail knowledge exchange, which is considered to be the most crucial knowledge management process in the OI-context.

2.2.1 Perspectives on Knowledge

Rooted in Penrose's (1959) theory of the firm and other research in the field of strategic management, Wernerfelt (1984) introduced the concept of the resource-based view, which assumes the possession of critical resources (i.e., tangible and intangible assets such as labor, capital, knowledge) lead to competitive advantages for the company holding these resources.[38] A few years later, when the resource-based view was already an established concept, Drucker (1993, p. 7) pointed out that knowledge is not only one of the traditional production factors (i.e., labor, capital, and land), but rather the most important and strategically significant resource for a company. By combining this idea with the resource-

---

[37] In the original article, the complete definition is written in italics and so highlighted here in a similar fashion.
[38] The resource-based view has become a key concept in strategic management, even though Wernerfelt's first article did not gain immediate recognition (see Wernerfelt 1995).

based view, the knowledge-based view evolved (see Grant 1996a, 1996b; Spender 1996). The basis of what exactly constitutes this resource has proven more elusive. A lot of people – from philosophers to economists – have tried to define it (cf. Nonaka and Takeuchi 1995, pp. 21ff.). Nevertheless, no broad consensus has been reached (Grant 1996b, p. 110). I will not compete in this area by providing my own definition. Nonetheless, some previous attempts to outline and differentiate knowledge in an economic context should be noted:

According to Davenport and Prusak (1998, p. 5), *"[k]nowledge is a fluid mix of framed experience, values, contextual information, and expert insight that provides a framework for evaluating and incorporating new experiences and information. It originates and is applied in the minds of knowers. In organizations, it often becomes embedded not only in documents or repositories but also in organizational routines, processes, practices, and norms."*

In addition to this definition, they drew a hierarchical distinction between knowledge, information and data – in the sense that data are the smallest units of this chain, which constitutes information as information constitutes knowledge. This causal relationship was also emphasized by Alavi and Leidner (2001, p. 109), who also pointed to the complexity of knowledge by introducing five alternative perspectives from which it might be viewed. Following their arguments, knowledge can be understood as a state of mind, an object, a process, an access to information, or a capability (cf. Alavi and Leidner 2001, pp. 109ff.). Depending on the position taken, the implications for knowledge management are different.[39]

A dominant and widely accepted taxonomy of knowledge draws on the work of Polanyi (1966) and is also implied by the above-mentioned definition by Davenport and Prusak: Knowledge can either be explicit (e.g., documents, repositories) or tacit[40] (e.g., routines, processes, practices, norms). Explicit knowledge can be coded and documented in writings or symbols. It is easy to articulate and to communicate and, thus, transferable from one person to another with reasonable effort. Thereby, explicit knowledge (except patents and copyright) shares various characteristics with public goods. Tacit knowledge, on the other hand, is very complex and difficult to reproduce in documents or databases. It is developed or arduously acquired by and stored within individuals, which makes it practically impossible to trade or transfer as a discrete, separate entity. The transfer of tacit knowledge is generally difficult, requires a lot of time and personal contact, and success is uncertain. It can only be revealed through application and acquired through observation and practice. All of these characteristics make tacit knowledge crucial for sustaining a competitive advantage. To some extent it is even more valuable than explicit knowledge because it is hard for

---

[39] In the context of open innovation, all five perspectives regarding knowledge are generally considered to be relevant. However, the process perspective, which focuses on the application of expertise and implies that knowledge management is concerned with knowledge flows and the creation, exchange, and distribution of knowledge, is especially close to the OI-concept.
[40] Tacit knowledge is related to some extent to the concept of "sticky information" (see Hippel 1994).

competitors to imitate. Another implication is that employees cannot be forced to share their tacit knowledge. Furthermore, it is hard to measure, monitor and, therefore, reward the contribution from an employee's tacit knowledge to a company's output (or punish the omission), which has important motivational effects as well as posing the risk of moral hazard (see Holmström 1979, 1982). (cf. Alavi and Leidner 2001, pp. 110ff.; Davenport and Probst 2002, pp. 70ff.; Grant 1996b, p. 111; Nonaka and Takeuchi 1995, pp. 8ff.; Nonaka et al. 2000, pp. 7f.; Osterloh and Frey 2000, p. 539; Krogh et al. 2000, pp. 82ff.; Winter 1987, pp. 170ff.)

The differentiation between explicit and tacit knowledge is very frequently used by scholars (see Foss et al. 2010; Kogut and Zander 1992; Oxley and Sampson 2004; Wasko and Faraj 2000) and R&D knowledge is assumed to have much in common with tacit knowledge (cf. Liu and Liu 2011, p. 983). Consequently, this differentiation will also be applied in this study.

2.2.2 Elements of Knowledge Management

Knowledge management is generally perceived as a process with various elements or phases. Similar to the OI-lifecycle (see chapter 2.1.3.3), academics basically agree on the underlying concept but the process delineation differs slightly with respect to the number and names of the single elements. For example, Alavi and Leidner (2001, p. 114) referred to knowledge creation, storage and retrieval, knowledge transfer, and the application of knowledge as basic elements of knowledge management. Foss et al. (2010, p. 457) suggested the key phases were creation, integration, sharing, and usage of knowledge, while other scholars named further steps (see e.g., Becerra-Fernandez and Sabherwal 2010; Cong et al. 2007; Lin and Lee 2004; Davenport and Prusak 1998). However, if all of the phases suggested in the literature are merged, a comprehensive knowledge management process with six steps evolves. As displayed in Figure 7, knowledge management is concerned with capturing relevant knowledge, creating or acquiring it, integrating the newly created/acquired knowledge into a given context and a proprietary knowledge base, using the knowledge, storing and retrieving it, and exchanging it.

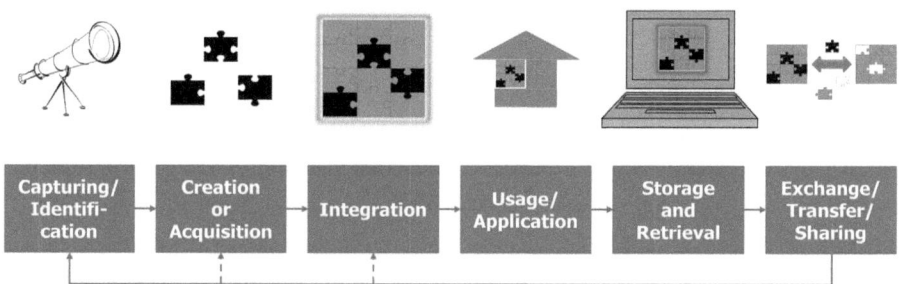

Figure 7: Knowledge Management Process[41]

The different phases of the knowledge management process are not necessarily sequential. Phases might be skipped or influence each other through feedback loops (e.g., knowledge exchange can lead to knowledge creation as in the case of open innovation). Knowledge management is not supposed to follow a strict process but to let knowledge flow and grow.

Looking at the phases from an OI-perspective, knowledge exchange in particular plays a crucial role[42] – even though all elements of knowledge management have a certain stake in open innovation. However, a comprehensive discussion of every element would extend the scope of this study exponentially. Therefore, in the following section, I will only focus on knowledge exchange.

2.2.3 Knowledge Exchange

A literature review conducted by Foss et al. (2010, p. 460) revealed that the phrases knowledge exchange, knowledge transfer, and knowledge sharing are often used interchangeably and, thus, can be considered synonyms. This finding confirms the view of Dixon (2000, p. 8), who asserted that many terms (e.g., exchange, transfer, disseminate) actually mean the same thing: They all describe the sharing of knowledge. However, Wang and Noe (2010, p. 117) dissented from this assertion by stating knowledge sharing is only a part of knowledge exchange and knowledge transfer. Applying their argument, knowledge exchange implies a reciprocal give-and-take relationship between individuals, i.e., people share and seeks knowledge. Knowledge transfer means a knowledge source shares and a recipient absorbs and uses the knowledge (cf. Davenport and Prusak 1998, p. 101). It has been typically used as term to describe knowledge flows between business units or organizations rather than individuals. This leads to the conclusion that, in a strict sense, the introduced terms (i.e., knowledge exchange, knowledge transfer, and knowledge sharing)

---

[41] Author's illustration
[42] Knowledge exchange is generally considered to be very important and the most essential element of knowledge management (cf. Bock and Kim 2002, p. 14).

are based on slightly different notions. However, the terms are used relatively interchangeably in the literature.

In this study, knowledge management and the process of exchanging, transferring, or sharing knowledge are viewed from an OI-perspective. Therefore, reciprocal knowledge flows between organizations and especially between individuals play a central role. The term knowledge exchange is considered to best describe this mutual relationship and is, therefore, the one applied most in this study. In this context, it is defined as a reciprocal knowledge sharing process[43], where both/all parties receive knowledge as well as share knowledge.[44] This working definition indicates that each individual shares his/her knowledge with the other involved individual(s), so that each party is a knowledge source and a knowledge recipient at the same time. The sum of all sharing efforts finally results in knowledge exchange (see Figure 8).

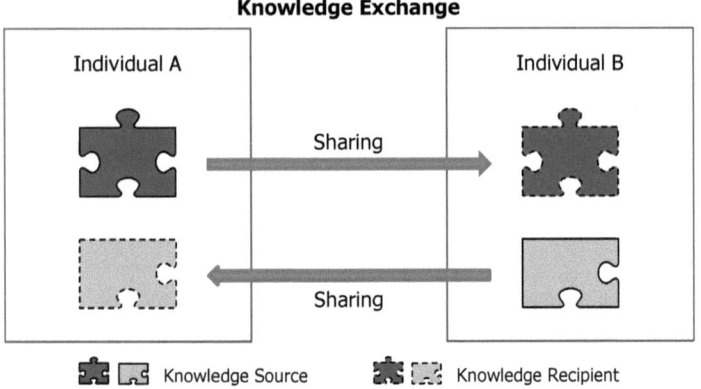

Figure 8: Concept of Knowledge Exchange[45]

As already alluded to, most of the literature does not differentiate among the existing terms related to knowledge flows and uses them interchangeably. However, knowledge sharing is by far the most frequently used term and dominates knowledge management literature. Therefore, the following explanations are often related to literature using this term.

---

[43] Finding a comprehensive definition of knowledge sharing is difficult because it depends to a large extent on the context (cf. Fengjie et al., pp. 278f.). However, many academics have attempted to define it. Hansen and Avital 2005, p. 6, say it refers to a situation whereby individuals voluntarily supply other individuals (internal or external to an organization) with his/her unique knowledge or experience.

[44] This definition was also stated on the front page of the online survey to ensure a common understanding of the key phrases in the questionnaire (see Appendix B).

[45] Illustration inspired by Cummings 2003, p. 40

2.2.3.1 Importance of Individuals for Knowledge Exchange

Following the knowledge-based view, the most important source of competitive advantage and a company's strongest value driver is knowledge, which inherently resides within individuals and, more precisely, within knowledgeable personnel. Consequently, the success of knowledge exchange is heavily dependent on employees' knowledge exchange efforts (cf. Bock et al. 2005, p. 88; Husted and Michailova 2010, p. 38). Since companies cannot force, but only encourage their employees to exchange knowledge (see Gibbert and Krause 2002; Osterloh and Frey 2000), employees are the ultimate decision-makers when it comes to exchanging or keeping their knowledge. They can freely decide when to exchange what with whom (cf. Husted and Michailova 2010, p. 40). For this reason, a reliable explanation of knowledge exchange processes should start at the micro-level and consider individual factors as key elements (cf. Foss et al. 2010, p. 459; Tohidinia and Mosakhani 2010, p. 613). Despite the obvious relevance of the employees' knowledge exchange behavior for the success of a company, relatively little is known about its determinants (cf. Bock and Kim 2002, p. 14). The literature on knowledge exchange neglects to build a micro-foundation and to formulate assumptions about individual actions, even though it would be important to obtain a better understanding about individual knowledge exchange behavior (cf. Foss et al. 2010, p. 465; Ho et al. 2009, pp. 1212f.).

2.2.3.2 Barriers and Drivers to Knowledge Exchange

If employees strategically decide about the point in time and extent of their knowledge exchange, goal tensions between the company and the employee might occur (cf. Husted and Michailova 2010, p. 40), which can lead to moral hazard – especially if the knowledge is tacit (see Holmström 1979, 1982). To encourage and facilitate knowledge exchange, companies should try to align employees' goals with the objectives of the organization and the project team (cf. Jewels and Ford 2006, p. 112). However, the possibly differing objectives of companies and employees are not the only barrier to knowledge exchange, as efforts to align the goals of both parties are not the only enabler. In the following, I give an overview of further barriers and drivers found in the literature and classify them according to the four dimensions suggested by Szulanski (1996, p. 30) and Cummings (2003, p. 9). Consequently, I distinguish between barriers and drivers related to the relational context, in which source and recipient are embedded, to characteristics of knowledge source and knowledge recipient, and to characteristics of the exchanged knowledge.

a) Characteristics of Relationship between Source and Recipient

Most of the barriers and drivers found in the literature can be attributed to the relational context (cf. Cummings and Teng 2003, p. 40). According to Cummings (2003, pp. 9ff.), five major relational barriers exist between the knowledge source and the knowledge recipient:

relationship distance, organizational distance, physical distance, institutional distance, and knowledge distance.

The first relational hurdle to knowledge exchange is the relationship distance between the knowledge-exchanging parties. It refers to the quality and extent of common past experience. One aspect of this hurdle and at the same time *"[...] the single most important precondition for knowledge exchange"* (Snowden 2000, p. 239) is trust, which can be defined in multiple ways and possess different dimensions (cf. Ford 2003, pp. 554ff. for an overview). Trust between parties is especially crucial where the knowledge exchange is very complex (i.e., the exchange of knowledge is risky and uncertain) and/or external control in form of contracts, rules, or policies assuring the proper behavior of all parties do not exist in an appropriate scope (cf. Ford 2003, p. 566).[46] Trust develops best from one's own past experiences with the partner (cf. Granovetter 1985, p. 490), personal contact, and the development of a healthy relationship with appropriate social ties (cf. Davenport and Prusak 1998, p. 97). In addition, Krogh et al. (2000, pp. 61f.) advised the creation of *"a sense of mutual dependence; make trustworthy behavior a part or performance reviews; and increase individual reliability by formulating a 'map' of expectations."* Besides trust, similarities between knowledge source and knowledge recipient can support the exchange of knowledge. This includes social and strategic similarities (see Child 1996; Peteraf and Shanley 1997), common thought processes and understanding of doing business (see Dougherty 1992), similar cultures and shared values (see Allen 1977), and a common language (cf. Davenport and Prusak 1998, p. 98). All these similarities are knowledge exchange drivers as they simplify communication, avoiding arduous relationships between parties (cf. Szulanski 1996, p. 32).

Organizational distance is a second relational barrier and arises if the source and the recipient are not settled in the same organization or unit. To overcome this, social ties should be established or strengthened. This is essential for the development of trust (see Parkhe 1998); it creates additional opportunities for knowledge exchange; and supports open communications between the parties (see Hansen 1999; Granovetter 1985). Furthermore, arrangements that structure and organize the knowledge exchange (e.g., contracts) can help reduce the organizational distance and also serve as enablers (see Baughn et al. 1997).

A third relational barrier to knowledge exchange is the physical distance between knowledge source and knowledge recipient (cf. Cummings and Teng 2003, p. 46). Face-to-face meetings and communication in general become more complicated and costly in terms of time and other expenses (see Nonaka and Takeuchi 1995). Therefore, physical distance generally reduces the effectiveness of knowledge flows (see Allen 1977; Lester and McCabe

---

[46] In addition, knowledge complexity and missing external control might correlate to a certain extent because it is hard to agree on and control for the concrete extent of knowledge exchange.

1993) and hampers knowledge exchange, since important issues are only addressed in face-to-face meetings (see Athanassiou and Nigh 2000) and complex or tacit knowledge is almost impossible to exchange without personal contact (cf. Davenport and Prusak 1998, p. 99). Furthermore, face-to-face meetings and regular personal contact facilitate the establishment of social ties and trust (cf. Davenport and Prusak 1998, pp. 97ff.). Therefore, investment in overcoming physical distance – even temporarily – can be valuable. One possibility is to agree on fixed times and places for meetings and knowledge exchange (cf. Davenport and Prusak 1998, p. 97).

A fourth relational barrier to knowledge exchange is the knowledge distance, which refers to the knowledge gap between source and recipient (cf. Cummings and Teng 2003, pp. 46f.). If the gap is too large, it is very likely the intellectually limited party can neither assimilate the received knowledge (cf. Hamel 1991, p. 97)[47] nor provide helpful knowledge to the other party. On the other hand, if the knowledge gap is too small, it might constrain the knowledge recipient's willingness to assimilate external knowledge because the recipient would be forced to dismiss some existing knowledge as the new knowledge replaces the old one. The existence of core-rigidity (see Leonard-Barton 1992) or the NIH-syndrome could additionally strengthen this effect.

A last relational obstacle to knowledge exchange is the institutional distance (see Kostova 1999). This hurdle occurs if the knowledge source and knowledge recipient are not located in the same country and are, therefore, embedded in different institutional contexts. This will probably entail differences with respect to regulatory (e.g., laws, rules), cognitive (e.g., schemas, frames), and normative (e.g., norms, values) aspects (cf. Kostova 1999, p. 314). These differences complicate the knowledge exchange. The institutional distance is therefore strongly related to the relational distance, as similarities between knowledge source and knowledge recipient are desirable to facilitate knowledge exchange.

b) Characteristics of Source

As already indicated, hurdles to knowledge exchange not only result from relational factors but also from the characteristics of the involved parties. Following Szulanski (1996, p. 31), one barrier related to the knowledge source could be an unwillingness to share crucial knowledge. The reasons can be manifold. Employees might be afraid to lose their knowledge, which could make them dispensable (cf. Eisfeldt 2009, p. 69). Another possible reason is that employees might simply be unwilling to put time and effort into the knowledge exchange. However, since an information asymmetry tends to exist between the source and the recipient (i.e., only the source knows what she/he knows) and it is hardly observable if the source is not able or willing to share knowledge, there is the danger of moral hazard,

---

[47] See also Lane and Lubatkin, 1998, who introduced the concept of relative absorptive capacity to address this issue.

which has to be addressed with appropriate instruments (see Holmström 1979, 1982). (cf. Cummings 2003, pp. 30f.)

A second source-related aspect results from the fact that *"[p]eople judge the information and knowledge they get in significant measure on the basis of who gives it to them."* (Davenport and Prusak 1998, p. 100) If a knowledge source is not perceived as reliable, the recipient might misjudge the value of the source's knowledge and challenge the input or even resist the absorption of valuable knowledge (cf. Szulanski 1996, p. 31). The reputation of the source therefore plays a crucial role in the knowledge exchange process. The credibility of the knowledge source is a key driver of knowledge exchange as it helps knowledge recipients recognize the value of the knowledge. (cf. Cummings 2003, p. 30)

Lastly, a knowledge source should be encouraged to positively influence the recipient's learning process to enable knowledge exchange (cf. Cummings 2003, p. 30). A knowledge source can facilitate knowledge exchange, for example, by expressing knowledge in an intelligible and interpretable manner. In addition, the source should only share relevant information so the recipient is not confronted with knowledge overkill and has to evaluate the relevance of the input. (cf. Alavi and Leidner 2001, p. 110)

c) Characteristics of Recipient

Besides the knowledge source, the knowledge recipient can significantly hamper or contribute to the knowledge exchange. According to Szulanski (1996, p. 31), a major barrier for the knowledge recipient is the NIH-syndrome. If the knowledge recipient is unwilling to accept external knowledge and absorb the source's input, an effective knowledge exchange becomes impossible.

A second critical aspect relates to the knowledge recipient's ability to exploit the external knowledge. If the recipient is willing to learn from the knowledge source but lacks in absorptive and retentive capacity,[48] any knowledge exchange effort is condemned to failure (cf. Szulanski 1996, p. 31). As absorptive capacity heavily depends on a recipient's stock of related knowledge, investments in internal R&D can extend the knowledge base and help to reduce this barrier. Therefore, internal R&D efforts can be drivers to knowledge exchange (see Cohen and Levinthal 1990).

The recipient's willingness and ability to embrace external knowledge is significantly affected by his/her learning culture. The characteristics of the learning culture can be an important barrier or driver to knowledge exchange. If routines and competencies to retain and nurture the received knowledge are established, freedom to invest significant time and other resources into knowledge exchange is granted, and mistakes are tolerated, the learning

---

[48] Retentive capacity means the recipient persistently applies the new knowledge, even though the initial integration of the received knowledge might be hard (cf. Szulanski 1996, p. 31).

culture is likely to positively influence the knowledge exchange. (cf. Cummings 2003, pp. 28f.)

A last aspect worth noting is the changing role of the knowledge recipient within the knowledge exchange process. As indicated in Figure 8, knowledge exchange is characterized by reciprocal sharing processes, i.e., knowledge exchange does not end when the recipient has obtained knowledge from the source. In fact, this is when the recipient becomes the new knowledge source and has to share knowledge with the former source, who is now the knowledge recipient. This role reversal might be an obstacle to knowledge exchange if the former knowledge recipient does not recognize this change and fails to perform the new tasks. (cf. Cummings 2003, p. 30)

d) Characteristics of Knowledge

Last but not least, the success of knowledge exchange depends on the characteristics of the exchanged knowledge. The general rule is: The more tacit and/or embedded (in people, tools, technologies, routines, tasks, etc.) the knowledge, the harder and more costly the knowledge exchange. Tacit knowledge can be more challenging to share because of "causal ambiguity" (see Lippman and Rumelt 1982). Causal ambiguity refers to the difficulty of defining (even ex post) the elements and factors (routines, tools, etc.) that have to be transferred to enable the recipient to properly replicate the knowledge (cf. Cummings 2003, pp. 20ff.; Foss et al. 2010, p. 468; Szulanski 1996, pp. 30f.).

2.3 Research Gap and Derivation of Research Questions

The review of the OI-related literature in chapter 2.1 indicates how broad the field of open innovation is, how many different research streams are related to OI-research, and the range of perspectives from which open innovation can be viewed. This makes it impossible to find an exhaustive definition of open innovation that is not too generic. In addition, the immense scope of this research field and affiliated streams is the reason gaps still exist, even though numerous scholars have already contributed to this field.

One of these gaps, which has been singled out by many academics and already discussed in chapter 2.1.3, refers to individuals engaged in open innovation (cf. Lichtenthaler 2011, p. 81; Vanhaverbeke 2006, pp. 206f.; Vrande et al. 2010, p. 230; West et al. 2006, pp. 287ff.):

> *"Although various research streams have contributed to research on Open Innovation and some of the different levels of analysis have been addressed in previous studies, the evolving debate is missing a key element: the people side of the equation."*
> *(Herzog 2008, p. 3)*

The rare articles relating to open innovation and the people side mainly focus on lead users or individuals engaged in open source projects and other OI-communities, rather than on the knowledge workers within an OI-embracing company.[49] The existence of this research gap is astonishing, since each OI-project starts with the efforts of employees or other individuals (cf. West et al. 2006, p. 287). Furthermore, the relevance of a micro-foundation had been highlighted by several scholars (see Coleman 1990; Gavetti 2005; Teece 2007). Empirical studies focusing on open innovation at the level of individuals in particular (e.g., employees) are missing in the literature and would make a great contribution to the research field (see Vrande et al. 2010).

> *"Therefore we hope that future research will explore Open Innovation at the individual or unit level [...]."* (Vanhaverbeke and Cloodt 2006, p. 279)

Assuming the R&D department of a company is mostly the place where companies' innovations begin and open innovation often takes place in the R&D environment, R&D employees play a pivotal role in open innovation (cf. Möslein 2009, p. 18; Möslein and Bansemir 2011, pp. 15f.). Consequently, the micro-foundation of OI-research should pay special attention to these individuals. The central task of R&D employees during the OI-process is to interact with external partners and to facilitate the in- and outflows of knowledge through knowledge exchange. By doing so, R&D employees lay the foundation for collaborative innovation. This makes companies heavily dependent on their R&D employees' knowledge exchange efforts. Nevertheless, this link has not yet been examined.[50] Consequently, the determinants of R&D employees' knowledge exchange behavior in OI-projects are wholly unknown. However, employees cannot be forced, but only encouraged to actively participate in open innovation. Therefore, companies need to know how to facilitate knowledge exchange in OI-projects and to motivate their R&D employees.

Internal barriers can be a major issue and might impede the successful implementation of an OI-approach. In this context, the role of employees' attitudes has been highlighted repeatedly (cf. Lichtenthaler 2011, p. 81). Accordingly, a substantial obstacle to open innovation and employees' willingness to be involved in OI-activities might result from the existence of the NIH-syndrome (see chapter 2.1.3.5), which reflects the negative attitude toward external ideas (cf. Clagett 1967, p. ii) and is a central component of the "want-barrier". To apply open innovation successfully, a company has to cope with all these internal barriers. Therefore, I formulated the following research questions (RQ):

RQ1: From an R&D perspective – What does open innovation mean and what aspects are especially important for knowledge exchange in OI-projects?

---

[49] For articles on open source projects and other OI-communities see for example Fleming and Waguespack 2007; Hars and Ou 2002; Henkel 2009; Jeppesen and Frederiksen 2006; and West and Lakhani 2008. For articles on lead users see for example Franke et al. 2006; Lüthje 2004; Schreier and Prügl 2008.
[50] Chatzoglou and Vraimaki 2009, p. 246, state that even knowledge exchange without the OI-aspect has not yet been adequately researched with respect to the empirical investigation of influencing factors.

RQ2: *Which factors determine the intention of R&D employees to exchange knowledge with external partners in OI-projects? Does a dominant factor exist?*

RQ3: *Which motivational factors can positively influence R&D employees' willingness to exchange their knowledge in OI-projects?*

2.4 Chapter Summary

Open innovation assumes internal and external knowledge is equally important to a company's innovation process and both should be used to optimize results. Furthermore, internal and external paths to market should be trod to optimally exploit company's innovations. These underlying ideas of open innovation are, however, not entirely new. OI-research is based on – and in line with – previous studies dealing with different external sources of innovation and their integration. Despite this background, when the OI-concept was introduced in 2003, it soon became a very popular research topic and publications on open innovation and affiliated research fields rocketed from then on. But although scholars could have researched open innovation on different levels, most of prior studies focused on the organizational level, neglecting the very relevant people side of the phenomenon. Scholars preferred to broaden an understanding of open innovation by researching aspects such as its adoption, the optimal degree of openness and its lifecycle. During this process, three archetypes of open innovation were identified (inbound, outbound, and coupled OI-processes), with coupled open innovation being the dominant type in praxis. In addition to the developments in OI-research, many affiliated streams emerged (e.g., crowdsourcing, mass customization) or advanced (e.g., open source, lead user concept). Furthermore, relations to other research fields such as organizational learning and knowledge management became apparent.

According to the knowledge-based view, knowledge is a company's strongest value driver and inherently resides within knowledgeable personnel. The exchange of knowledge (inflows and outflows of knowledge) plays a central role in the OI-concept and is understood in this thesis as a reciprocal knowledge sharing process, where both/all parties receive knowledge as well as share it. Consequently, R&D employees engaged in OI-activities are knowledge sources and receivers at the same time. This illustrates R&D employees' immense role in OI-projects and related knowledge exchange processes and, moreover, the importance of conducting research on open innovation and knowledge exchange at both individual and employee levels. Since companies cannot force but only encourage their R&D employees to exchange knowledge, it is important for companies to obtain a better understanding of employees' motives for exchanging knowledge in OI-projects and to recognize potential barriers. This argument is reinforced by the fact that knowledge in an R&D context is often tacit. The exchange of tacit knowledge is comparably difficult; it is hard to measure and to monitor, which could affect motivation and pose the risk of moral hazard. Since companies cannot control the exchange of their employees' tacit knowledge in OI-projects, they are heavily dependent on employees' willingness and on their abilities to exchange their knowledge in OI-projects. This suggests the barriers identified in the context of open innovation (impacting employees' willingness, ability, and perceived desirability) are closely linked with the barriers to knowledge exchange (related to relational context, characteristics of the knowledge sources and recipients, and characteristics of the exchanged knowledge). In particular, employees' knowledge source and knowledge recipient characteristics are affected by their willingness and abilities to exchange knowledge. Figure 9 gives an overview of the central aspects of this study and of the related research questions.

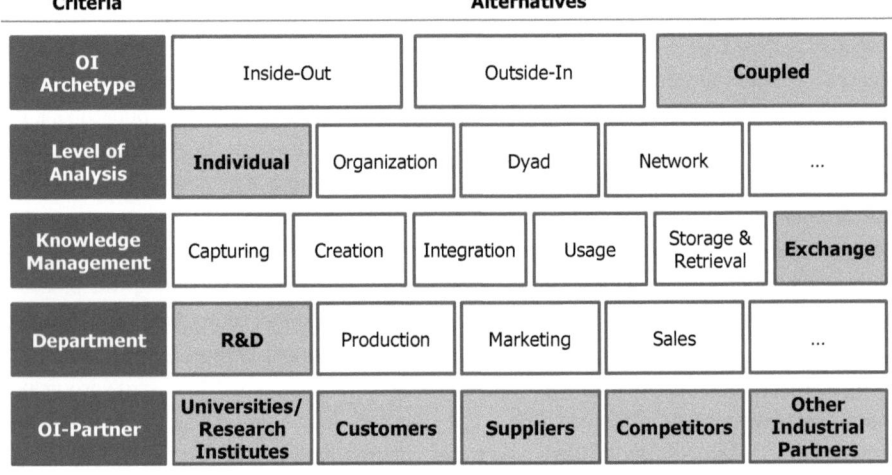

Figure 9: Research Focus[51]

---

[51] Illustration inspired by Wecht 2006, p. 5

3 Theoretical Foundation

This chapter focuses on introducing the theories consulted to derive hypotheses for the research model of this study and to answer the formulated research questions. The theory of planned behavior (TPB) – a frequently used theory with great predictive validity (see Armitage and Conner 2001) – builds the theoretical foundation of this study. Therefore, the first sub-chapter discusses the components of this theory and the underlying relationships. The second sub-chapter concentrates on publications where researchers have applied the TPB to examine individuals' knowledge exchange behavior.[52] This literature review mainly aims to identify motivational factors that have an impact on employees' willingness to exchange knowledge in OI-projects. The identified motivational factors and their hypothesized influence on employees' willingness are discussed in the last sub-chapter.

3.1 Theory of Planned Behavior

The TPB was introduced by Ajzen (1985) and is an extension of Fishbein and Ajzen's (1975; 1980) theory of reasoned action (TRA). Both theories aim to explain human behavior and assume individuals' intentions to be the most important influencing factor. Intention, in turn, is determined by a person's attitude toward a behavior (A) and the subjective norm (SN). The difference between both theories relates to an individual's control over his/her behavior. In contrast to the TRA, the TPB assumes not every action is under volitional control and that perceived behavioral control (PBC) directly influences the intention as well as the behavior of individuals (see Figure 10).

As displayed in Figure 11, the components of the TPB can be related to the behavioral barriers identified in chapter 2.1.3.5: If individuals do not want to behave in a specific manner ("want-barrier"), it is often accounted for by their negative attitude toward this behavior, i.e., attitude is related to individuals' willingness. The "shall-barrier" can be considered the result of the subjective norm, i.e., the subjective norm is related to the social desirability of the behavior. Lastly, the missing ability to behave in a specific manner ("can-barrier") might be linked to the lack of perceived behavioral control, i.e., perceived behavioral control is related to individuals' ability.

---

[52] Since the knowledge exchange behavior of individuals had not yet been researched in the context of open innovation, I had to base my thesis on the existing literature.

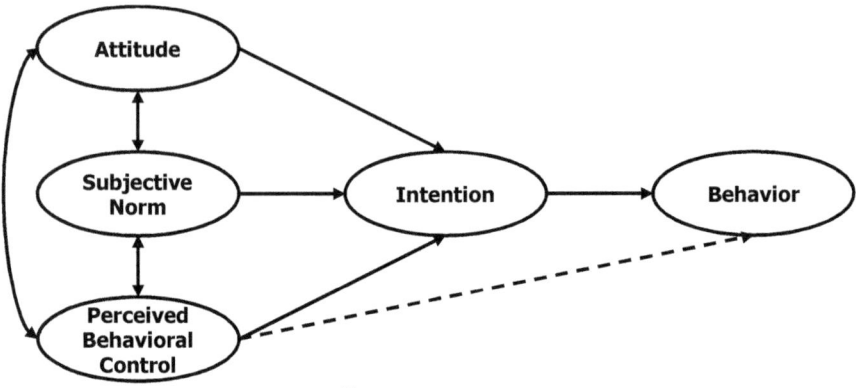

Figure 10: Theory of Planned Behavior[53]

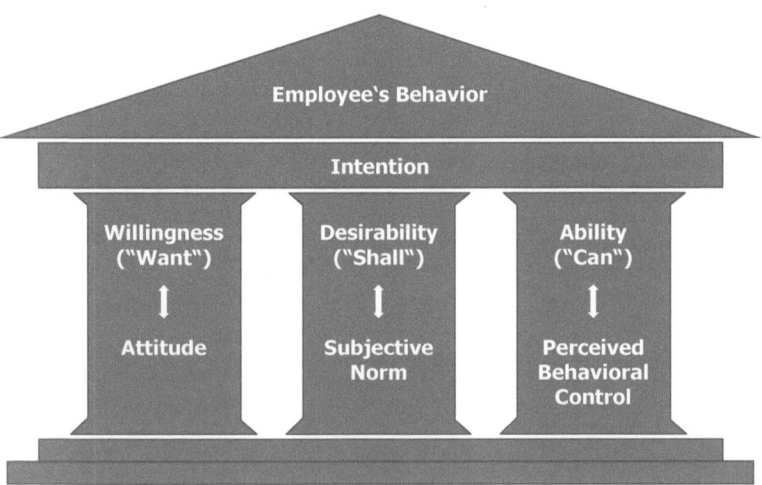

Figure 11: Theory of Planned Behavior and Sources of Behavioral Barriers[54]

Again, the TPB was coined by Ajzen and developed from joint research with Fishbein. Therefore, unless cited, the following exposition of the TPB (i.e., the forthcoming sub-chapters) is based on the related standard works of these two scholars (see Ajzen 1985, 1991; Ajzen and Fishbein 1975, 1980).

---

[53] Illustration: Ajzen 1991, p. 182
[54] Author's illustration

3.1.1 Attitude

In the context of TRA and TPB, attitude *"[...] refers to the degree to which a person has a favorable or unfavorable evaluation or appraisal of the behavior in question."* (Ajzen 1991, p. 188) It develops from people's salient beliefs[55] about the behavior. These salient beliefs assign certain outcomes to the performance of behavior. Thus, attitude can be considered as a function of the strength of behavioral beliefs (i.e., subjective likelihood) that the performance of a behavior will lead to certain outcomes (b_i) and the subjective evaluations of these anticipated outcomes (e_i):[56]

$$A \propto \sum_{i=1}^{n} b_i e_i$$

If certain behavior is believed to lead mostly to positive outcomes, people hold a favorable attitude toward carrying it out. On the contrary, if a mode of behavior is anticipated to have mainly negative effects, the attitude will be unfavorable. Therefore, people learn to prefer behavior with positive consequences and to avoid behavior with negative outcomes.

3.1.2 Subjective Norm

In the context of TRA and TPB, the subjective norm *"[...] refers to the perceived social pressure to perform or not to perform the behavior"* (Ajzen 1991, p. 188) and develops from people's normative beliefs. These normative beliefs represent the perceived likelihood that important others (i.e., individuals or groups such as family, friends, colleagues, supervisors, etc.) approve or disapprove of the behavior in question. However, this perception does not necessarily reflect the actual opinion of the referents. Subjective norm can be considered a function of the strength of normative beliefs (i.e., subjective likelihood) that the behavior is approved or disapproved by important others (n_i) and the motivation to comply with the referents (m_i):

$$SN \propto \sum_{i=1}^{n} n_i m_i$$

If a person assumes the behavior to be supported or desired by the referents and the motivation to comply with these important individuals or groups is high, the perceived social pressure (i.e., subjective norm) is also high.

---

[55] It is assumed a person can hold multiple beliefs about a given behavior. However, only few (i.e., salient beliefs) actually determine an individual's attitude at any given point in time.
[56] Bandura noted that the outcome of performing a certain action is not always in line with the previously assigned target outcome: *"Many actions are performed in the belief that they will bring about a desired outcome, but they actually produce outcomes that were neither intended nor wanted."* (Bandura 2003, p. 3)

3.1.3 Perceived Behavioral Control

The construct of perceived behavioral control was added to the TRA to address the issue that not every behavior is under complete volitional control, i.e., people cannot always freely decide whether or not to behave in a certain way. The TPB, which resulted from this extension, considers potential constraints on the behavior in question (cf. Minbaeva and Pedersen 2010, p. 204) and, thus, is able to predict behavior that is not completely under volitional control. In the context of TPB, perceived behavioral control *"[...] refers to the perceived ease or difficulty of performing the behavior and it is assumed to reflect past experience as well as anticipated impediments and obstacles."* (Ajzen 1991, p. 188) It develops from people's control beliefs. These control beliefs represent the assumed presence or absence of resources, opportunities, and personal abilities necessary for performing the behavior in question. They evolve from past experience and second-hand information (e.g., other people's experiences). Therefore, perceived behavioral control can be considered a function of the strength of control beliefs (i.e., subjective likelihood) that required resources, opportunities, and capabilities for performing the behavior are present (c_i) and the perceived power of these control factors to facilitate the performance of the behavior (p_i):

$$\text{PBC} \propto \sum_{i=1}^{n} c_i p_i$$

If a person believes he/she has all of the relevant resources, opportunities, and abilities for performing the behavior and does not anticipate a lot of obstacles, the performance of the behavior is felt to be easy, i.e., perceived behavioral control is high.

After Ajzen had introduced the construct of perceived behavioral control in the context of the TPB, other scholars claimed it actually comprises two independent variables: self-efficacy and perceived behavioral control over behavior (see Armitage and Conner 1999a, 1999b; Manstead and Eekelen 1998; Terry and O'Leary 1995). Ajzen (2002b) picked up this criticism and agreed perceived behavioral control is indeed composed of the two elements:[57]

> *"Recent research has demonstrated that the overarching concept of perceived behavioral control, as commonly assessed, is comprised of two components: self-efficacy (dealing largely with the ease or difficulty of performing a behavior) and controllability (the extent to which performance is up to the actor). [...] This view of the control component in the theory of planned behavior implies that measures of perceived behavioral control should contain items that assess self-efficacy as well as controllability."* (Ajzen 2002b, p. 680)

---

[57] In addition, Ajzen noted the term "perceived behavior control" might have been misleading. "Perceived control over performance of a behavior" (Ajzen 2002b, p. 668) would be less ambiguous. Nevertheless, I have continued to use the term "perceived behavioral control" because it is used throughout the literature.

Perceived controllability reflects a person's evaluation of how much personal control he/she has over performing the behavior in question. Self-efficacy is a concept coined by Bandura (see e.g., Bandura 1977, 1982, 2003) and describes a person's judgment of his/her own ability to execute a certain action.[58] However, *"[...] perceived self-efficacy is concerned not with the number of skills you have, but with what you believe you can do with what you have under a variety of circumstances."* (Bandura 2003, p. 37) A person perceiving a low degree of self-efficacy will assume the performance of the behavior in question to be difficult (cf. Bandura 2003, p. 127), although her/she might possesses all of the relevant capabilities to execute the behavior, i.e., self-doubt can over-rule skills to such an extent that highly talented people might not efficiently use their capabilities (cf. Bandura 2003, p. 37). In contrast, if the sense of perceived self-efficacy is strong, obstacles might even spur people to put greater effort into performing the behavior (cf. Bandura 1982, p. 123).

3.1.4 Intention

An individual's intention (I) to perform a certain behavior is a central factor in TRA and TPB. It reflects a person's motivation to engage in a given behavior and the subjective likelihood that the person will perform it (cf. Armitage and Conner 1999a, p. 74). Thus, intentions serve as *"[...] indications of how hard people are willing to try, of how much of an effort they are planning to exert, in order to perform the behavior."* (Ajzen 1991, p. 181) Consequently, intentions can be considered direct antecedents of individual behavior (cf. Ajzen 2002a, p. 1). The TPB assumes that behavioral intention develops from individual's attitude, subjective norm, and perceived behavioral control (see Figure 10). According to Ajzen (1991, p. 188), the general rule is that the more favorable an individual's attitude toward performing the behavior, the more encouraging subjective norm with respect to the behavior and the greater the individual's perceived behavioral control, the stronger will be individual's intention to perform the behavior in question. The relative importance of the three factors in predicting intention might vary across situations and behaviors but some empirical findings suggest individual's attitude toward performing the behavior often play an important role (see e.g., Ajzen and Madden 1986; Ajzen 1991; Armitage and Conner 1999a; Blue 1995). This view is also supported by Fishbein and Ajzen (1975, p. 288), emphasizing the strong relationship between attitude and intention and by Lichtenthalter (2011, p. 81), underlining the relevance of individuals' attitudes.[59]

---

[58] Bandura 2003, pp. 11ff pointed out that self-efficacy is not the same as related concepts, such as self-esteem or locus of control even though the terms are often confused. For a differentiation between self-efficacy and self-esteem in the organizational context see Gardner and Pierce 1998.

[59] Lichtenthaler 2011 argues attitude can contribute significantly to the micro-foundation of OI-research. Furthermore, he says major barriers to open innovation might evolve from individuals' attitudes (e.g., NIH-syndrome). This argument finds support in Clagett's definition of the NIH-syndrome as a negative attitude of a technical organization toward external ideas and innovations (cf. Clagett 1967, p. ii).

3.1.5 Behavior

The behavior of individuals and its prediction is the main concern of TRA and TPB. The TPB suggests behavior that is completely under volitional control develops from an individual's intention to perform the behavior. In principle, the stronger the intention to perform a certain behavior, the more likely is the performance of this behavior. However, behavioral intention can only lead to the intended behavior if the behavior in question is under volitional control. Factors such as violation of physical integrity, lack of willpower, underestimation of the behavior's difficulty and/or unanticipated consequences, and natural disasters can limit individuals' control and cause a gap between their intention and the actual behavior (cf. Ajzen and Fishbein 1980, p. 48; Ajzen 2002b, pp. 665f.). If individuals cannot freely decide about performing or not performing the behavior, perceived behavioral control becomes an additional important variable that directly influences the behavior (see Figure 10). However, the relative importance of intention and perceived behavioral control in the prediction of behavior depends on the situation and the behavior (see Kuo and Young 2008a, 2008b for empirical investigations of this issue).

To investigate a certain behavior, it is important to define it in terms of target, action, context, and time (TACT). Furthermore, all other constructs of the TPB have to be aligned accordingly (i.e., intention, attitude, subjective norm, and perceived behavioral control), so that the definitions and formulations of all variables are compatible with each other (cf. Ajzen 2002a, pp. 2f.). The relevance of this alignment arises from the fact that the strength of relationship between the constructs – especially between intention and behavior – is heavily dependent on the corresponding measurements. If these are not compatible with respect to the four TACT elements, an intention-behavior gap is likely. The more specific the behavior of interest; the more the other constructs are defined; and the more compatible the measures, the better the predictive power of the TPB.

Another aspect significantly determining the strength of the intention-behavior relationship is the stability of intentions (cf. Ajzen and Fishbein 1980, pp. 47ff., 2005, p. 188). Intentions can change, particularly at the individuals' level, where the relationship between intention and behavior is relatively unstable. However, stability is required to predict behavior correctly. To ensure maximum stability, intention and behavior should be assessed without a great time lag. In addition, intention and behavior must be measured independently to avoid literal inconsistency (cf. Ajzen and Fishbein 2005, pp. 189f.).

As indicated through my research questions (see chapter 2.3), this study aims to investigate R&D employees' knowledge exchange with external partners in OI-projects. Due to the focus on the individual, the relationship between intention and behavior is not expected to be very stable. This means the assessment of both factors must be conducted without any great time lag. Furthermore, the existence of literal inconsistency would entail asking R&D

employees about their intentions and colleagues or supervisors about the actual behavior of these same employees. The combination of both requirements would make it a very complex and time-consuming – if not impossible – task for companies to identify the matching couples and to deliver all of the relevant data in time. Therefore, I have decided to exclude the behavior construct from my study and to focus on the prediction of R&D employees' intention to exchange knowledge with external partners in OI-projects.

3.2 TPB and Knowledge Exchange – A Literature Review

After deciding to hone my research on the TPB, I searched for articles where the TPB (or the TRA) had already been applied in a context similar to my research field (i.e., knowledge exchange in OI-projects). This literature review had two objectives: The first and main goal was to identify variables that would presumably significantly influence individuals' attitude toward exchanging their knowledge in OI-projects. This objective was based on the special predictive role of attitude (see chapter 3.1.4) and the associated research question about motivational factors influencing R&D employees' willingness to exchange their knowledge with external partners (see chapter 2.3). The second objective was to find established measures that could later be used for the operationalization of the constructs included in my research model. However, I could not find any literature combining the TRA or TPB with research on open innovation or knowledge exchange in OI-projects. Therefore, I focused the search on articles investigating the knowledge exchange of individuals by way of the TRA or TPB and used this as a proxy for the original literature review.

To identify relevant studies, I employed the EBSCOHost Research Database and Google Scholar[60] and combined the phrases "knowledge exchange", "knowledge sharing", and "knowledge transfer" with the search terms "theory of planned behavior" and "theory of reasoned action".[61] Studies without a clear focus on knowledge exchange were ignored. Articles that neither had considered predictors of attitude in their research, nor stated their applied questionnaire items were also excluded from detailed inspection. After filtering the search results accordingly, I went through the reference list of all the remaining articles to identify further important studies. At the end of this process, I had compiled a list of 24 relevant articles (see Table 1). The majority of these studies based their research on the TPB. However, in ten cases TRA was the underlying theory. In 17 of the 24 articles, predictors of attitude were integrated into the research model. These publications greatly contributed to the identification of attitude-predicting motivational factors and to the

---

[60] See http://search.ebscohost.com and http://scholar.google.de/ (both last accessed on May 30, 2013)
[61] I also used British English search terms (i.e., behaviour) and the abbreviations TPB and TRA so as not to miss relevant articles. Consequently, 15 search runs were conducted in total: The three phrases: "knowledge exchange", "knowledge sharing", and "knowledge transfer" were each combined with the five phrases: "theory of planned behavior", "theory of planned behaviour" (British English), "theory of reasoned action", "TPB", and "TRA".

selection of relevant constructs for my study. 19 of the 24 articles stated the applied questionnaire items and, therefore, played a key part in the later operationalization phase (see chapter 4.4.1).

Table 1: Literature Review

	Source	Applied Theory	Predictor of Attitude	Questionnaire Items
1	Bock and Kim (2002)	TRA	Included	–
2	Bock (2005)	TRA	Included	Included
3	Chatzoglou and Vraimaki (2009)	TPB	–	Included
4	Chow and Chan (2008)	TRA	Included	Included
5	Erden et al. (2012)	TPB	Included	Included
6	Ho et al. (2009)	TRA	(Included)[†]	–
7	Huang et al. (2008)	TRA	Included	Included
8	Jeon et al. (2011)	TPB	Included	Included
9	Jewels and Ford (2006)	TPB	–	Included
10	Kuo and Young (2008a)	TPB	–	Included
11	Kuo and Young (2008b)	TPB	–	Included
12	Kwok and Gao (2005)	TRA	Included	Included
13	Lin (2007a)	TRA	Included	Included
14	Lin and Lee (2004)	TPB	–	Included
15	Minbaeva and Pedersen (2010)	TPB	Included	Included
16	Ryu et al. (2003)	TPB	–	Included
17	So and Bolloju (2005)	TPB	–	Included
18	Teh et al. (2010)	TPB	Included	–
19	Teh and Yong (2011)	TRA	Included	Included
20	Tohidinia and Mosakhani (2010)	TPB	Included	Included
21	Wu and Wei (2010)	TPB	Included	–
22	Xie (2009)	TPB	Included	–
23	Yang and Lai (2011)	TPA	Included	Included
24	Zhang and Ng (2012)	TRA	Included	Included

† This study applies game theory instead of structural equation modeling, so that predictors of attitude are stated, but the predictive power is not assessed for each individual factor.

Table 2 gives an overview of the 17 publications with predictors of individual attitudes in their research model. All of the studies apart from two were conducted in Asian countries. The sample size ranges from 70 to 531 responses. For the derivation of factors likely to predict individuals' attitude toward exchanging their knowledge, the researchers often relied on economic and social exchange theory (see e.g., Bock and Kim 2002; Zhang and Ng 2012), which are two of the principle theories to explain social interactions.[62] Since knowledge exchange is a type of social interaction between individuals, the application of these two theories suggests itself (cf. Bock and Kim 2002, p. 15). Most of the derived factors likely to influence attitude were hypothesized to be positively related to attitude (i.e., the greater/stronger ... the more favorable the attitude to exchange knowledge). Indeed, the majority of hypotheses were supported by the data. With respect to the hypotheses that were not supported, it is striking that more than 25% of these are related to rewards.

---

[62] For further reading on economic and social exchange theory, refer to Blau 1964; Emerson 1976; Homans 1961; Kelley and Thibaut 1978.

Table 2: Articles with Predictors of Attitude

Source	Sample	Predictor of Attitude	Hypothesis	Result*
Bock and Kim (2002)	N = 467 Four large companies, Korea	Expected associations Expected contribution Rewards	+ + +	+ + −
Bock (2005)	N = 154 27 companies across 16 industries, Korea	Reciprocity Rewards Sense of self-worth	+ + +	+ − o
Chow and Chan (2008)	N = 190 Managers, Hong Kong, China	Shared goals Social network Social trust	+ + +	+ + o
Erden et al. (2012)	N = 531 Online community members, Korea	Community munificence	+	+
Ho et al. (2009)	N = 70 Three large high-tech companies, Taiwan	Expected associations Expected contribution Level of understanding Rewards Self-esteem Cost of sharing Self-interest	+ + + + + − −	Game theory approach
Huang et al. (2008)	N = 159 MBA students, China	Image Reciprocity Rewards Sense of self-worth Codification effort Loss of knowledge-power	+ + + + − −	o o + + o −
Jeon et al. (2011)	N = 282 Four large high-tech companies, Korea	Enjoyment in helping Image Need for affiliation Reciprocity	+ + + +	+ + + +
Kwok and Gao (2005)	N = 75 Students, Hong Kong, China[63]	Absorptive capacity Channel richness Extrinsic motivation	+ + −	o + o
Lin (2007a)	N = 172 50 companies across 15 industries, Taiwan	Enjoyment in helping Knowledge self-efficacy Reciprocity Rewards	+ + + +	+ + + o
Minbaeva and Pedersen (2010)	N = 470 Two large companies, Denmark	Rewards	+	−
Teh et al. (2010)	N = 301 Students, Malaysia	Internet self-efficacy	+	+
Teh and Yong (2011)	N = 116 Three IT-companies, Malaysia	In-role behavior Sense of self-worth	+ +	+ +
Tohidinia and Mosakhani (2010)	N = 502 50 oil-companies, Iran	Reciprocity Rewards Self-efficacy	+ + +	+ o +

---

[63] This information was derived from the statement that the data were collected in an information systems department. Since only one of the two authors' works in such a department, it was assumed that his university and country are the origin of the data.

Wu and Wei (2010)	N = 150 Students, Taiwan	Enjoyment in helping	+	+
		Expected contribution	+	+
		Expected relationship	+	o
		Disincentives	+	+
		Positive reinforcement	+	o
		Expected loss	−	−
		Sharing interference	−	o
Xie (2009)	N = 322[64] 13 industries, China	Extrinsic motivators	+	o
		Intrinsic motivators	+	+
		Org. commitment	+	+
		Org. climate	+	o
Yang and Lai (2011)	N = 219 Wikipedia members	Information quality	+	+
		System quality	+	+
Zhang and Ng (2012)	N = 231 Construction workers, Hong Kong, China	Enhanced relationship	+	o
		Knowledge feedback	+	+
		Knowledge self-efficacy	+	+
		Reduced workload	+	o
		Rewards	+	o
		Losing face	−	−

+ Positive relationship hypothesized / significant positive effect
− Negative relationship hypothesized / significant negative effect
o No significant effect
* Results with minimum significance level p < 0.05

After gathering all attitude-influencing factors from the 17 publications, I decided to add only the most relevant factors to my research model so as not to extend the scope of this study. Assuming the frequency of application would indicate the relevance of the factors to some degree; I included four of the most frequently used constructs: enjoyment in helping, sense of self-worth (which is very closely related to the construct of expected contribution), reciprocity and rewards.[65] This selection received further support from Davenport and Prusak (1998, pp. 31ff.), who stated that at least three kinds of payments exist in knowledge markets: altruism (i.e., enjoyment in helping), reciprocity, and repute.

3.3 Hypotheses and Research Model

In order to build the foundation for an empirical analysis, the research questions formulated in chapter 2.3 had to be merged with the underlying theories of this study, i.e., the assumed relationships were converted into verifiable hypotheses. Together with the constructs, these hypotheses constituted the final research model.

3.3.1 Theory of Planned Behavior

The TPB is at the core of my research model. Therefore, the first three hypotheses are derived from underlying assumptions of this theory. As already mentioned in chapter 3.1.4,

---

[64] The information given in the article's abstract and in the article itself is contradictory. The abstract states N = 322. In the article N = 320.
[65] Self-efficacy was also repeatedly used as a predictive factor in the 17 studies. As self-efficacy was already dealt with in connection with perceived behavior control (see chapter 3.1.3), it is not discussed further.

the TPB generally considers individual's attitude, subjective norm, and perceived behavioral control as antecedents of behavioral intention. It assumes all three factors to be positively related to individual's intentions, even though the relative predictive power of the factors might vary across situations and behaviors (see Ajzen 1985, 1991). This set of relationships and the related assumptions have been examined in various contexts (including in the context of knowledge exchange) where the TPB received considerable support from multiple empirical studies (see e.g., Jeon et al. 2011; Lin and Lee 2004; Minbaeva and Pedersen 2010; Ryu et al. 2003; Tohidinia and Mosakhani 2010). There is no reason to believe the TBP would not apply to knowledge exchange in OI-projects, so I have defined the following hypotheses for my study:

H1: *R&D employees' attitude toward exchanging their knowledge with external partners in OI-projects has a positive impact on their intention to exchange knowledge with external partners in OI-projects.*

H2: *The subjective norm concerning knowledge exchange with external partners in OI-projects has a positive impact on R&D employees' intention to exchange their knowledge with external partners in OI-projects.*

H3: *R&D employees' perceived behavioral control over their knowledge exchange with external partners in OI-projects has a positive impact on their intention to exchange knowledge with external partners in OI-projects.*

3.3.2 Enjoyment in Helping

Enjoyment in helping is related to pro-social behavior and the concept of altruism (cf. Jeon et al. 2011, p. 256). Altruism is a kind of payment in knowledge markets and reflects people's motivation to exchange knowledge without expecting more than a "thank you" in return (cf. Davenport and Prusak 1998, pp. 31ff.). However, this understanding already indicates that people are hardly motivated by pure or absolute altruism. According to Smith (1981, p. 23), even seemingly complete altruistic behavior incorporates some selfish elements, such as the unconscious hope for feel-good rewards or intrinsic satisfaction from helping others, for example. Therefore, he argued for the motivational effect of relative altruism and defined it as *"[...] an aspect of human motivation that is present to the degree that the individual derives intrinsic satisfaction or psychic rewards from attempting to optimize the intrinsic satisfaction of one or more other persons without the conscious expectation of participating in an exchange relationship whereby those 'others' would be obligated to make similar/related satisfaction optimization efforts in return."* (Smith 1981, p. 23) Following this definition, altruism and the enjoyment in helping belongs among the intrinsic motivators, which are generally important for knowledge exchange and considered superior to extrinsic motivators when it comes to the generation and exchange of tacit knowledge (cf. Osterloh and Frey 2000, p. 540).[66] The importance of altruism and the enjoyment derived from helping with respect to an individual's knowledge exchange behavior received empirical support from the study of Wasko and Faraj (2000, pp. 164ff.), who showed that enjoyment in helping does indeed motivate people to exchange knowledge. Furthermore, several researchers have examined the predictive power of enjoyment in helping with respect to individuals' attitudes to exchanging knowledge and found a significantly positive relationship between both variables (cf. Table 2 and see Jeon et al. 2011; Lin 2007a; Wu and Wei 2010). Therefore, I have defined the following hypothesis for my study:

H4: *Enjoyment in helping has a positive impact on R&D employees' attitude toward exchanging their knowledge with external partners in OI-projects.*

3.3.3 Sense of Self-Worth

A person's sense of self-worth is part of his/her overall self-concept (see Kinch 1963, 1973) and is related to self-esteem (cf. Gecas 1982, pp. 4f.). According to Bandura (2003, p. 11), self-esteem reflects the judgment of self-worth. A person can derive his/her sense of self-worth from different fields (work, family life, etc.), but the extent may vary depending on the field, i.e., a person might take great pride from his/her achievements at work but not so much from his/her family life and vice versa (cf. Bandura 2003, p. 12). In the context of an organization and with respect to knowledge exchange, sense of self-worth *"[...] captures the extent to which employees see themselves as providing value to their organizations through their knowledge sharing."* (Bock et al. 2005, p. 91) Following Cabrera and Cabrera (2002, p. 695), employees are more willing to exchange knowledge if they expect to make a considerable contribution and so generate value for their company. Feedback regarding their contribution is therefore an important control mechanism (cf. Kinch 1973, pp. 55, 77). Positive feedback can encourage employees' knowledge exchange, while negative feedback can help to control the quality of employees' contributions (cf. Cabrera and Cabrera 2002, p. 699).

The definition of sense of self-worth given above is compatible with the idea of organization-based self-esteem (see Pierce et al. 1989), which *"[...] reflects an employee's evaluation of his or her personal adequacy and worthiness as an organizational member."* (Gardner and Pierce 1998, p. 50) This was hypothesized and shown to influence employees' attitudes (cf. Gardner and Pierce 1998, pp. 54ff.). Several other researchers have also examined the predictive power of sense of self-worth with respect to individuals' attitudes to exchange knowledge and mostly found a significant positive relationship between both variables

---

[66] For more details about the distinction between intrinsic and extrinsic motivation, see Ryan and Deci 2000; Staw 1976.

(cf. Table 2 and see Bock et al. 2005; Huang et al. 2008; Teh and Yong 2011). As a result, I have defined the following hypothesis for my study:

H5: *Sense of self-worth (in an organizational context) has a positive impact on R&D employees' attitude toward exchanging their knowledge with external partners in OI-projects.*

3.3.4 Reciprocity

Similar to altruism, reciprocity is considered a form of payment in knowledge markets (cf. Davenport and Prusak 1998, pp. 31f.).[67] It represents a pattern of mutual exchange, dependence and indebtedness between two or more parties (cf. Gouldner 1960, p. 170; Ipe 2003, p. 346; Lin 2007a, p. 139; Molm 1997, p. 20), and entails that each party has rights, but also obligations, which result from a history of previous interactions between the parties (cf. Gouldner 1960, pp. 169, 171). However, reciprocity is not simply either present or absent. Following Gouldner (1960, p. 164), the total presence (i.e., exchanged benefits are identical in form or equal in value, cf. Gouldner 1960, p. 172) or absence (i.e., one party receives benefits but does not return anything) of reciprocity seldom occurs in social relations. Rather, it presents the two extremes of a continuum. Consequently, an intermediate level of reciprocity – where all parties receive some benefits but one party invests a little more than the other parties – is more common. However, each person who provides benefits to others in social relations has to accept a minimum of cost (investment cost; direct cost; opportunity cost), since time and effort at least have to be devoted to the relationship (cf. Blau 1964, p. 101).[68] Therefore, people try to minimize costs and maximize benefits in order to achieve a maximum outcome (cf. Molm 1997, pp. 13ff.). As a consequence, they will only engage in social relations if the expected outcome of their behavior attains a certain level (cf. Kelley and Thibaut 1978, pp. 8ff.).

Besides its function as a form of currency in knowledge markets, reciprocity also serves as an initiator of social interactions and a stabilizer of social relations (cf. Gouldner 1960, p. 176). The stabilizing function of reciprocity is found in the nature of exchange and in the time delay between initial benefits and repayment. A person providing any kind of benefit to another person obliges that individual, who in turn must reciprocate in some way to discharge obligations (cf. Blau 1964, p. 89). Since only in rare cases, compensation directly follows the initial benefit and creditors tend to maintain relationships with outstanding obligations, a stable social relationship evolves (cf. Gouldner 1960, pp. 174f.).

---

[67] Unless stated, the following explanations assume a positive reciprocity (i.e., the exchange or return of benefits). Nevertheless, most arguments also apply to negative reciprocity (i.e., the exchange or return of harm), which does exist (see Constant et al. 1994) and is historically "*[...] the most important expression of homeomorphic reciprocity [...] best exemplified by the lex talionis.*" (Gouldner 1960, p. 172; The underlined words are highlighted through italics in the original source.).

[68] In a reciprocal relationship, each party will incur costs because each party will provide a certain benefit.

The above discussion indicates reciprocity is closely related to social exchange theory (see Blau 1964; Emerson 1976; Homans 1961; Kelley and Thibaut 1978). According to Blau (1964, p. 91), "*'[s]ocial exchange' [...] refers to voluntary actions of individuals that are motivated by the returns they are expected to bring and typically do in fact bring from others.*" In contrast to strict economic exchange, where the quantity and value of resulting obligations are already negotiated and stipulated in the beginning, social exchange creates unspecified obligations (cf. Blau 1964, p. 93). Exchange between employees and their company is often a blend of economic and social exchange (cf. Organ and Konovsky 1989, p. 162). However, knowledge exchange between employees and external partners within OI-projects can primarily be considered as social exchange, since the resulting obligations for the individual employee cannot be foreseen precisely at the beginning of an OI-project.

The importance and motivational power of reciprocity with respect to individual's knowledge exchange behavior received empirical support in the study of Wasko and Faraj (2000, pp. 165ff.). However, the study did not confirm that people expected direct reciprocity as noted in the social exchange theory (cf. Homans 1961, p. 2), but rather a generalized form of reciprocal behavior (cf. Molm 1997, pp. 20f.). This means people do not necessarily expect to get back knowledge exactly from the person to whom they gave their knowledge. However, they do expect to receive knowledge in return from someone. Several other researchers have also examined the predictive power of reciprocity with respect to individuals' attitudes toward exchanging knowledge and mostly found a significant positive relationship between both variables (cf. Table 2 and see Bock et al. 2005; Huang et al. 2008; Jeon et al. 2011; Lin 2007a; Tohidinia and Mosakhani 2010). Therefore, I have defined the following hypothesis for my study:

H6: *Reciprocity has a positive impact on R&D employees' attitude toward exchanging their knowledge with external partners in OI-projects.*

3.3.5 Rewards

As already mentioned, exchange theory indicates the behavior of individuals is guided by their dominant objectives of maximizing benefits and minimizing costs (see Molm 1997). Consequently, they prefer to engage in activities with positive outcomes, i.e., those where benefits exceed the costs (cf. Homans 1961, p. 61). This implies people also expect to receive rewards for participating in interactions with others (cf. Kelley and Thibaut 1978, pp. 8ff.), which is why Davenport and Prusak (1998, pp. 32, 47f.) pointed out the need to reward knowledge exchange.

> "*Rewards could range from monetary incentives such as bonuses to non-monetary awards such as dinner gift certificates to awards such as praise and public recognition that do not have a monetary equivalent value.*" (Bartol and Srivastava 2002, p. 66)

This definition depicts the ability to differentiate rewards in at least two ways: They may be monetary or non-monetary; tangible or intangible. Overall, rewards are considered extrinsic motivators (cf. Bartol and Srivastava 2002, p. 66).[69] The key characteristic of extrinsic motivators is that they are not directly related to the behavior that should be stimulated (cf. Osterloh and Frey 2000, p. 539; Ryan and Deci 2000, p. 62), i.e., *"[...] extrinsically motivated behavior is controlled by incentives that are not part of the activity [...]"* (Eisenberger and Cameron 1996, p. 1154). Consequently, extrinsically motivated behavior is very outcome-driven and aims to obtain rewards and/or avoid penalties (cf. Kelley and Thibaut 1978, pp. 8ff.; Kowal and Fortier 1999, p. 357).

Several researchers examined the predictive power of rewards with respect to individuals' attitudes to exchange knowledge (see Bock and Kim 2002; Bock et al. 2005; Ho et al. 2009; Huang et al. 2008; Lin 2007a; Minbaeva and Pedersen 2010; Tohidinia and Mosakhani 2010; Zhang and Ng 2012). Based on the exchange theory and people's objective to maximize benefits (see Kelley and Thibaut 1978; Molm 1997), all of them anticipated a positive relationship between both variables (see Table 2). Surprisingly, most of the studies either could not find a significant relationship at all or found a significant negative effect on individuals' attitudes to exchange knowledge. Herzberg's motivation-hygiene theory (see Herzberg 1968, 1974) in combination with the operationalization of the reward construct attempt to explain this observation.[70] Following his theory, a differentiation between factors leading to job satisfaction (i.e., motivators) and factors leading to job dissatisfaction (i.e., hygiene factors) is imperative.[71] The presence of motivators creates satisfaction. However, their absence does not lead to dissatisfaction but only to no satisfaction. In contrast, the absence of hygiene factors induces dissatisfaction but their presence does not lead to satisfaction but only to no dissatisfaction. Considering all of the above-mentioned studies that provide an overview of the applied questionnaire items, it becomes clear that the operationalization of the reward construct draws mainly on elements that are hygiene factors rather than motivators (e.g., salary, bonus, job security). This implies the reward construct measures dissatisfaction rather than satisfaction. Consequently, rewards seem to have no effect or to even impede the formation of a positive attitude toward knowledge exchange. In addition, knowledge exchange in OI-projects can scarcely

---

[69] According to Deci and Ryan 1985, extrinsic motivators will have a negative impact on intrinsic motivation because extrinsic motivators such as monetary rewards undermine the self-determination of the individual, which is the basis of intrinsic motivation (cf. Bartol and Srivastava 2002, p. 66). If companies want their employees to have a high level of creativity and interest in their work, incentive systems might have a negative impact (cf. Eisenberger and Cameron 1996, p. 1153). However, this view is not shared by all scholars in this field (see Eisenberger and Cameron 1996).
[70] The motivation-hygiene theory is also known as the "two-factor theory" (cf. Herzberg 1974, p. 18).
[71] According to Herzberg et al. 1959 and Herzberg 1968, 1974, achievement, recognition of achievement, work itself, responsibility, advancement, and growth are all examples of motivators. Company policy and administration, supervision, salary, status, and security are examples of hygiene factors.

be attributed to a single person and rewarded directly.[72] Therefore, I assume in this specific context that rewards negatively influence employees' attitude toward exchanging their knowledge in OI-projects. Accordingly, I have defined the following hypothesis for my study:

> *H7: Rewards have a negative impact on R&D employees' attitude toward exchanging their knowledge with external partners in OI-projects.*

3.4 Chapter Summary

The theory of planned behavior, which builds the theoretical foundation of this study, assumes that any behavior is best predicted by the intention to perform it and by the perceived behavioral control. In turn, the intention is determined by three factors: attitude, subjective norm, and perceived behavioral control (see Figure 10). In the past, the TPB has been applied in various research fields, such as in knowledge management, but never in the OI-context. Since no empirical studies have yet combined the TPB with research on open innovation or knowledge exchange in OI-projects, literature about knowledge exchange was considered the closest possible proxy to identify factors with an impact on attitude toward knowledge exchange in OI-projects. A review of studies related to knowledge exchange and the application of the TPB or TRA revealed four factors that were repeatedly hypothesized to significantly influence individuals' attitudes toward exchanging their knowledge.

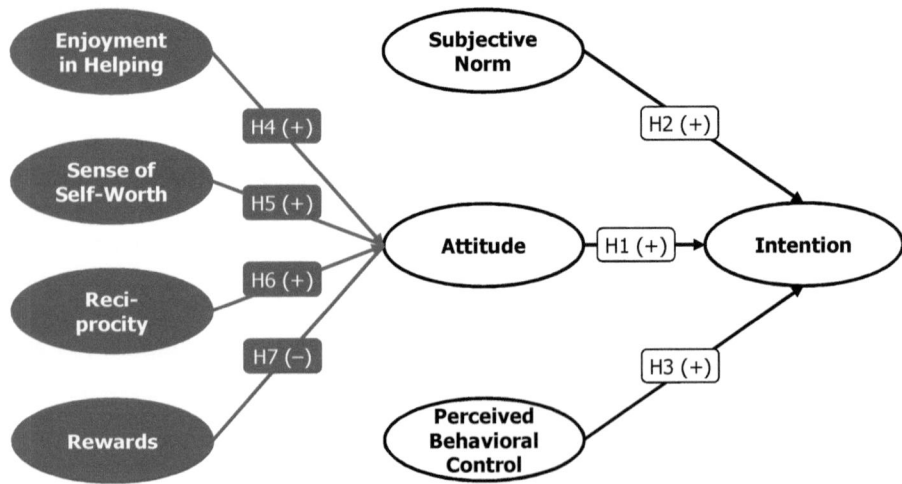

Figure 12: Research Model[73]

---

[72] This is especially true if the exchanged knowledge is tacit (see Osterloh and Frey 2000).
[73] Author's illustration

Chapter Summary

Based on the TPB and this literature review, seven hypotheses were finally derived and brought together in one research model (see Figure 12). This research model served as the basis for the quantitative part of this study. As shown in Figure 12, all of the relationships in the research model – apart from the connection between rewards and attitude (H7) – were anticipated to be positive.

4 Research Design and Operationalization

In this chapter, the underlying research approach and the different phases of the empirical part of this study are described. Due to the focus of my thesis (see chapter 1), it was necessary to align the research design to the context of knowledge exchange in OI-projects and to the aspired level of research – R&D employees with experience in OI-projects and collaboration with external partners. In the following sub-chapters, I explain the research approach and reasoning behind the company selection. Thereafter, details on the qualitative pre-study and the quantitative main study are provided.

4.1 Research Approach

Generally, research in social science can be divided into qualitative and quantitative methods (cf. Walliman 2006, pp. 36f.). Qualitative methods such as interviews (cf. Bernard 2000, pp. 190ff.), case studies and group discussions are inductive research approaches. They are commonly applied in research fields where few or no previous studies exist. Therefore, the derivation of well-grounded hypotheses proves to be difficult. They aim for explorative theory building and can contribute greatly in emerging research fields. Their disadvantages include their limited objectivity and the particularization of their results i.e., these are not representative across an entire population. Quantitative methods such as surveys, on the other hand, involve deductive research to test rather than build theory. Hypotheses and a theoretical framework are required, which can be tested and verified by the data. Due to the application of standardized measures and usually high sample sizes, quantitative methods are more objective and a generalization of results might be possible (see Bryman 2008).

Due to the different qualities and characteristics of qualitative and quantitative methods, it is often considered beneficial to mix both techniques so as to combine their advantages; to try to overcome their weaknesses; and to view the research topic from different perspectives (cf. DeCuir-Gunby 2008, pp. 125f.). This is particularly apparent in the increasing tendency to apply the mixed-method approach, which can be found in the literature (cf. Bryman 2010, p. 30 and see for example Grote 2010; Harkness 2006; Janzik 2012; Pettersson et al. 2009; Torres 2006; Schweisfurth 2013). The rationale behind the application of mixed methods is diverse. Greene et al. (1989, p. 259) classified the purpose of combining qualitative and quantitative methods into five groups displayed in Figure 13:

- **INITIATION** seeks the discovery of paradox and contradiction, the recasting of questions/results from one method with questions/results from the other method.
- **COMPLEMENTARITY** seeks elaboration, enhancement, illustration, clarification of the results from one method with results from the other method.
- **TRIANGULATION** seeks convergence, corroboration, correspondence of results from the different methods.
- **EXPANSION** seeks to extend the breadth and range of inquiry by using different methods for different inquiry components.
- **DEVELOPMENT** seeks to use results from one method to help develop or inform the other method (regarding sampling, implementation, measurement decisions).

Figure 13: Purposes for Mixed-Method Approach[74]

In consideration of the advantages and disadvantages of the different research methods, I split my research into three phases and adopted a mixed-method approach. The first phase was a qualitative pre-study in which I conducted 12 interviews with R&D managers. The purpose of all 12 interviews was to better understand open innovation in a specific organizational context, to assess the OI-awareness among the R&D staff and to ensure all participating companies had a comparable understanding of open innovation. The interviews were also intended to finalize the questionnaire[75] and, thereby, to prepare the main part and second phase of my research – an online survey (cf. McQueen and Knussen 2002, pp. 13f.). This survey addressed R&D employees and was intended to test the research model and related hypotheses. Since the target group was multinational, the survey was provided in German and English. The responses were analyzed and presentations were prepared based on the outcomes.[76] The findings from the online survey formed the basis for the third and last phase. In the context of three follow-up group discussions, they were discussed with a group of scholars and two groups of R&D representatives from two participating companies. These group discussions were primarily supposed to explain surprising and unexpected results that diverged with those found in the literature review. In addition, the discussions

---

[74] Illustration: cf. Greene et al. 1989, p. 259
[75] When the interviews were conducted, a research model had already been derived from the literature review and a questionnaire using existing items and constructs was compiled. Since OI-research lacks empirical contributions, it was not possible to find appropriate constructs to investigate governance mechanisms within OI-projects. Therefore, I sought to develop a new construct by using the interviews as the basis for research on regulations and conditions that are – from an employee's point of view – necessary conditions in OI-projects. Despite all efforts to identify governance elements that would apply to different kinds of OI-projects, it became clear that OI-projects are too heterogeneous and the development of an appropriate construct would extend the scope of this thesis. Therefore, I decided to confine this aspect to the addition of an open question to the questionnaire, which might be informational and provide a starting point for future research.
[76] Each of the participating companies received an individualized presentation. The presentations were also used as basis for the follow-up group discussions.

were intended to validate the results from the survey and to check whether the relationships thrown up by the analysis of the survey data were a true reflection of reality.

4.2 Selection of Companies

Company selection was led by some preliminary considerations. It had to be assumed that corporate culture and other company-specific characteristics might have an influence on the research topic and, thus, an effect on employees' responses and the results of the study. Such company-specific aspects are hard to measure and to compare. Furthermore, it was unlikely every relevant aspect would be included appropriately in the study. For the sake of minimizing this bias, it was considered beneficial to focus the research on a small selection of companies and involve a substantial number of R&D employees from each company instead of approaching a lot of companies and involving only one or two of their employees.

A chosen company had to fulfill three main criteria: Given the research topic, it was essential the company had experience of open innovation. More importantly, it had to be aware of its OI-experience, i.e., it was crucial that the company knew about the concept of open innovation and was able to identify OI-projects and initiatives it had embarked upon. Public communication of applying open innovation was used as a proxy for the company's OI-awareness. More precisely, a company had to state the words "open innovation" in its official company website, in annual reports, and/or in press releases to be considered as a relevant candidate.[77] The second criterion related to the size of the R&D department. A potential candidate had to have an R&D department with a significant number of employees. This was important because the intended research approach (see chapter 4.1) and the preliminary considerations explicated at the beginning of this sub-chapter focused on establishing strong ties with a selection of companies and convincing these companies to involve around 40 OI-experienced R&D employees in the online survey.[78] Lastly, it was considered beneficial to focus the search for participating companies on corporations headquartered in German-speaking countries (i.e., Germany, Austria, and Switzerland). The purpose of this criterion was to facilitate personal contact, on-site presentations, and the establishment of strong ties with the companies. However, although this criterion simplified the communication with participating companies, it also diminished the basic population for the search process. So as not to unnecessarily reduce the population further the search for companies was not restricted to a specific industry category but it was confined to the broad spectrum of manufacturing industries (services were excluded).

---

[77] As the focus was on companies headquartered in German-speaking countries, German equivalents for open innovation (e.g., "offene Innovation") were also acceptable.
[78] The goal of collecting 40 responses from every company originated in the research model (see Figure 12) and the minimum requirement for a data analysis method discussed in chapter 4.4.5.

Two publicly available company lists were consulted as a starting point to initiate the search for relevant companies. The first one named "udaba.de - Die Unternehmensdatenbank im Internet"[79] contained around 1,000 firms that are well-known in Germany and at least operate in German-speaking countries (even if they are not headquartered in these). The Forbes list of "Global 2000 Leading Companies"[80] was the second source. Despite the high number of initial companies, only 21 fulfilled all of the criteria and were identified as relevant participants. Consequently, only 21 companies out of almost 3,000 openly communicated the application of the OI-approach, were manufacturers with a significantly large R&D department, and were headquartered in a German-speaking country.[81] These companies were contacted and asked to participate in interviews and/or the online survey. In 18 cases, a contact person from R&D, Innovation Management or the Product Management division could be identified and was directly contacted via e-mail.[82] The other three companies had to be contacted via the official contact form on their website.

In total, six companies agreed to participate in my study. However, due to time constraints two of them could only offer one or two employees, confined to the online survey. For reasons outlined at the beginning of this sub-chapter, these two companies were not included in the analysis. The remaining four companies were manufacturers with global businesses, headquartered in Germany, active in the B2B[83] market, and operating in the fields of chemistry, automation, and steel treatment.[84] In each of the four companies, one person was assigned as a central contact person.

4.3 Qualitative Pre-Study (Interviews)

R&D managers from all four companies contributed to the pre-study; though the number and duration of the interviews varied. The purpose of the interviews was to explore how the R&D managers define open innovation; why their companies follow the OI-approach; which phases a company has to pass; and which requirements have to be fulfilled in order to

---

[79] See www.udaba.de (last accessed: June 12, 2012)
[80] See www.forbes.com/global2000/ (last accessed: June 12, 2012)
[81] The first step was to exclude service providers and redundant companies from the udaba-list. After excluding them, 494 out of around 1,000 companies remained. In a second step, service providers, companies already identified in the udaba-list and companies not headquartered in Germany, Austria, or Switzerland were eliminated from the Forbes list. After the elimination, 33 of 2,000 companies remained. In total, 527 companies were kept in the pool of potential participants and so were further examined. In a third step, the official company website, annual reports, and/or press releases from these 527 companies were searched for their headquarters and for the phrase "open innovation" (or a German equivalent). If open innovation was mentioned and the company was headquartered in Germany, Austria, or Switzerland, the enterprise was marked as a relevant participant.
[82] The contact persons were mainly the heads of department.
[83] The term Business-to-Business is often used in the field of e-commerce and means a company (= business) offers its products or services to other businesses (i.e., B2B), rather than to end-consumers (i.e., B2C) (cf. Gabler Verlag 2004, p. 572).
[84] Since I assured the companies of absolute anonymity, I cannot go into further detail about the selected companies.

benefit from open innovation. In the following, information about data collection, the sample and data analysis method are presented.

4.3.1 Data Collection and Sample

In total, I conducted 12 interviews. Eight of them were semi-structured telephone interviews (cf. Bernard 2000, p. 191) lasting approximately one hour each. Another semi-structured interview was conducted in a 30-minutes face-to-face meeting. The last three were unstructured interviews (cf. Bernard 2000, p. 191). Two of them lasted approximately 15 minutes and were conducted by telephone, the other one was a 30-minute, face-to-face interview. Table 3 gives an overview of all the interviews.[85]

The interviewees were all higher level R&D managers with experience in open innovation. The interviews with representatives from companies A and B were conducted in March and April 2012, the other two in June and July 2012. I interviewed the central contact person of company A, who appointed five of his colleagues for further interviews. At company B, the contact person gave me an interview during an on-site visit and named three other R&D managers. At companies C and D, only the contact persons were interviewed.

Table 3: Overview of Interviews

	Structure	Communication	Duration	Company	Code
Interview 1	Semi-structured	Face-to-face	30 minutes	A	A1
Interview 2	Semi-structured	Telephone	60 minutes	A	A2
Interview 3	Semi-structured	Telephone	60 minutes	A	A3
Interview 4	Semi-structured	Telephone	60 minutes	A	A4
Interview 5	Semi-structured	Telephone	60 minutes	A	A5
Interview 6	Semi-structured	Telephone	60 minutes	A	A6
Interview 7	Semi-structured	Telephone	60 minutes	B	B1
Interview 8	Semi-structured	Telephone	60 minutes	B	B2
Interview 9	Semi-structured	Telephone	60 minutes	B	B3
Interview 10	Unstructured	Face-to-face	30 minutes	B	–
Interview 11	Unstructured	Telephone	15 minutes	C	–
Interview 12	Unstructured	Telephone	15 minutes	D	–

To prepare for the semi-structured interviews, I had formulated guidelines (see Appendix A). The questions were structured along three topics. The first group of questions dealt with open innovation in general. Interviewees were asked to give their own definition of open innovation and to highlight its meaning for their company. They were additionally asked if employees in the company knew about the OI-concept and if internal synonyms for open innovation existed. The second group of questions referred to OI-projects. For example, I inquired about the duration and process of an OI-project, the strength and weaknesses, and the typical OI-partners. The last part of the interview guidelines focused on basic

---

[85] The interview codes (see the last column of Table 3) are only relevant for the analysis of the interviews (see chapter 5).

conditions and requirements for OI-projects. In this section, I attempted to find out those factors that facilitated open innovation and those that might throw up barriers. I asked about legal issues and intellectual property and what arrangements are made to prevent damages. Furthermore, I tried to find out about employees' motivations and attitudes regarding open innovation and asked about any particular measures that prepare employees for their assignments in OI-projects. However, due to time constraints it was not always possible to cover all of the questions in every interview.

4.3.2 Method of Data Analysis

In preparation for the analysis, I transcribed all nine semi-structured interviews according to the suggested rules of Dresing and Pehl (2011), which resulted in 130 pages of text. Together with the memos from three unstructured interviews, these transcripts became the basis for the analysis. Since the interviews were not conducted with the purpose of theory building, but rather to aggregate the interviewees' opinions by reducing the material to the very central points and structuring the answers, a qualitative content analysis was conducted (see Kuckartz 2007; Mayring 2008).

As a first step, I read the printed transcripts thoroughly, marked all meaningful passages and derived rough categories. The categories were strongly related to the main topics and questions mandated by the interview guidelines. After finishing the first step on paper, I transferred the interviews to MaxQDA[86] and used the identified categories to develop a first coding system. Thereafter, I worked through the interviews once again and coded meaningful passages by applying the developed coding system. At the same time, I searched for repeatedly mentioned aspects and curious statements to refine the coding system and code the interviews accordingly. This procedure also helped separate substantial segments and statements from irrelevant interview passages. After all the interviews had gone through the coding process, I prepared one list per sub-code with all retrieved segments, i.e., all statements assigned to the same sub-code were collated in a list. Each list of accumulated quotes was then printed and read again thoroughly. The most interesting parts were marked and keywords were highlighted, which laid the foundation for the final interpretation of the data. Some of the marked passages were directly extracted and served as direct quotes. Other passages were consolidated, rephrased, and indirectly cited. The keywords helped to structure all of the quotes and contributions into a coherent narrative. As a final step, all direct and indirect quotes were interpreted and related to each other and – where possible – to the existing literature. As the interviews were originally in German, the first draft of the analysis was written in German and then translated to English. (see Gläser and Laudel 2004)

---

[86] MaxQDA is a software package for the analysis of qualitative date. For this analysis MaxQDA 10 was employed (see MaxQDA 2010).

4.4 Quantitative Study (Online Survey)

R&D employees from all four companies participated in the online survey, even though the number of responses differed. The main purpose of the questionnaire was to test the research model and related hypotheses. In the following, the operationalization of constructs and the pre-test procedure are described. Furthermore, I give a detailed overview of how data were collected, cleansed, and prepared for the analysis. Lastly, the final sample and data analysis method is described.

4.4.1 Operationalization of Constructs

Most of the measures applied in my thesis have been used in other studies before and showed a respectable degree of reliability and validity. Table 4 presents the TPB constructs and items employed. Table 5 gives an overview of the motivational constructs and items used in my study.[87]

Since the survey was conducted online, questions could be defined as mandatory, implying that the questionnaire could not be finished without having answered all of the mandatory questions. In order to assure a high usability of the responses, all model-related items were defined as mandatory. Furthermore, most of the questions relating to the control variables – except age, gender, and an open question regarding OI-requirements – were defined as mandatory.

As mentioned in chapter 4.1, a German and an English version of the online survey was prepared. The original items were all in English and so had to be translated into German. In order to ensure consistency, backwards translation was employed, i.e., I translated the questionnaire from English into German and three academic researchers translated it back into English. Subsequently, the original items were compared with the three English versions. Finally, discrepancies were resolved. (see Mullen 1995; Singh 1995)

---

[87] Items of control variables (e.g., OI-experience, OI-partners) are not included in the overview. For the complete questionnaire see Appendix B.

Table 4: Operationalization of Theory of Planned Behavior Constructs

Construct	Code	Item
Attitude	A1[†]	My knowledge exchange with external partners in OI-projects is ... *(very harmful/very beneficial)*
	A2	My knowledge exchange with external partners in OI-projects is a ... experience. *(very unpleasant/very pleasant)*
	A3	My knowledge exchange with external partners in OI-projects is ... to me. *(very worthless/very valuable)*
	A4	My knowledge exchange with external partners in OI-projects is a ... move. *(very unwise/very wise)*
	A5	Overall, my knowledge exchange with external partners in OI-projects is ... *(very bad/very good)*
Subjective Norm		**Normative Beliefs**
	SNn_1	My CEO wants me to exchange knowledge with external partners in OI-projects.
	SNn_2	My immediate supervisor wants me to exchange knowledge with external partners in OI-projects.
	SNn_3	My colleagues want me to exchange knowledge with external partners in OI-projects.
		Motivation to Comply
	SNm_1	Generally speaking, I try to follow the CEO's policy and intention.
	SNm_2	Generally speaking, I accept and carry out my immediate supervisor's decision even though it is different from mine.
	SNm_3	Generally speaking, I respect and put in practice my colleagues' decision.
Perceived Behavioral Control		**Perceived Controllability**
	PBC1	Whether or not I exchange knowledge with external partners in OI-projects is entirely up to me.
	PBC2	I have full personal control over exchanging knowledge with external partners in OI-projects.
		Perceived Self-Efficacy
	PBC3	If it is entirely up to me, I am confident that I am able to exchange knowledge with external partners in OI-projects.
	PBC4	I believe I have the ability to exchange knowledge with external partners in OI-projects.
	PBC5	I am capable of exchanging knowledge with external partners in OI-projects.
	PBC6[†]	For me exchanging knowledge with external partners in OI-projects is ... *(very difficult/very easy)*
Intention		**Intention to Exchange Documented Knowledge (Intention_doc)**
	I1	I will exchange work reports and official documents with external partners in future OI-projects.
	I2	I will exchange manuals, methodologies, and models with external partners in future OI-projects.
		Intention to Exchange Undocumented Knowledge (Intention_undoc)
	I3	I will exchange experience or know-how from work with external partners in future OI-projects.
	I4	I will provide my know-where or know-whom at the request of external partners in OI-projects.
	I5	I will exchange my expertise from my education or training with external partners in future OI-projects.

[†] These items were deleted after factor analyses (see chapter 6.4.1)

Table 5: Operationalization of Motivational Constructs

Construct	Code	Item
Enjoyment in Helping	JOY1	I enjoy exchanging knowledge with external partners in OI-projects.
	JOY2	I enjoy helping others by exchanging knowledge with external partners in OI-projects.
	JOY3	It feels good to help someone else by exchanging knowledge with external partners in OI-projects.
Sense of Self-Worth		My knowledge exchange with external partners in OI-projects ...
	SW1	... helps other members in my organization to solve problems.
	SW2[†]	... creates new business opportunities for my organization.
	SW3	... improves work processes in my organization.
	SW4	... increases productivity in my organization.
	SW5	... helps my organization to achieve its performance objectives.
Reciprocity		When I exchange knowledge with external partners in OI-projects ...
	RP1[†]	... I believe that I will get knowledge for giving knowledge.
	RP2	... I expect somebody to respond when I'm in need.
	RP3	... I expect to get back knowledge when I need it.
	RP4	... I believe that my queries for knowledge will be answered in future.
Rewards		When I exchange knowledge with external partners in OI-projects it is important to me ...
	Literature Based	
	REW1	... to get better work assignments.
	REW2	... to be promoted.
	REW3	... to get a higher salary.
	REW4	... to get a higher bonus.
	REW5[†]	... to increase my job security.
	Pre-test Based	
	REW6	... to enhance my reputation.
	REW7	... to build a network.
	REW8	... to increase my knowledge.
	REW9[†]	... to improve my job performance.
	REW10	... to add value for my company.

† Items were deleted after factor analyses (see chapter 6.4.1)

4.4.1.1 Attitude

For the operationalization of R&D employees' attitude toward knowledge exchange with external partners in OI-projects, I followed the instructions of Ajzen (2002a) on how to construct a TPB questionnaire. I decided to apply direct measures (cf. Ajzen 2002a, pp. 4f.). Hence, the attitude construct comprised five items and was reflectively[88] measured on a 5-point Likert scale[89] (very harmful/very beneficial; very unpleasant/very pleasant; very worthless/very valuable; very unwise/very wise; very bad/very good)[90]. On the basis of

---

[88] At this point, I will not go into detail regarding the differences between reflective and formative measures. For further explanation see Diamantopoulos and Winklhofer 2001; Diamantopoulos and Siguaw 2006; Jarvis et al. 2003, 2005; and Petter et al. 2007
[89] A five-point scale was chosen for this and for all other constructs with a bipolar scale because five response categories leave enough – but not too much – room for differentiation and can therefore be considered optimal (cf. Bradburn et al. 2004, p. 330; Dillman et al. 2009, p. 137).
[90] Following Bradburn et al. 2004, p. 329, the response categories of all scales in this study were listed from negative to positive and from lowest to highest.

Chatzoglou and Vraimaki's (2009) scale for measuring bank employees' knowledge sharing behavior, the items' wording was adapted to the context of knowledge exchange in OI-projects. Several studies on knowledge exchange have used this measure or a slightly modified version of it (see Bock et al. 2005; Erden et al. 2012; Huang et al. 2008; Jeon et al. 2011; Lin 2007a; Tohidinia and Mosakhani 2010; Zhang and Ng 2012).

4.4.1.2 Subjective Norm

The instructions on how to create a TPB questionnaire from Ajzen (2002a) was also the basis for the operationalization of subjective norm, which reflects the perceived social pressure to exchange knowledge with external partners in OI-projects. Instead of direct measures, I opted for belief-based measures, where the strength of normative beliefs (n_i) and the employees' motivation to comply (m_i) those perceived social demands are measured and multiplied (cf. Ajzen and Madden 1986, p. 462). In the end, subjective norm is calculated according to the following formula (cf. Ajzen 1991, p. 195, 2002a, p. 12):

$$SN \propto \sum_{i=1}^{n} n_i m_i$$

In order to avoid possible scaling issues, it is recommended to normalize the calculated scores of subjective norm (cf. Bock et al. 2005, p. 95; Jeon et al. 2011, p. 258). This was done in this study according to the approach of Bailey and Pearson (1983).

With regard to the formulation and number of items, I adapted the measures from Bock et al. (2005), which have also been used by Huang et al. (2008) and in an extended version by Jeon et al. (2011). In the study of Bock et al. (2005), subjective norm is the sum of the perceived social pressures created by the CEO, the immediate supervisor, and colleagues. Consequently, subjective norm was operationalized using three items comprising two questions each.[91] The items' wording was adapted to the OI-context and a 5-point Likert scale was used. For the three questions to measure the strength of normative beliefs, the scale ranged from "very unlikely" to "very likely". For the three questions to measure the motivation to comply, the scale ranged from "strongly disagree" to "strongly agree".

---

[91] Three questions related to the strength of normative beliefs about the CEO, the immediate supervisor, and colleagues. Three more related to the associated motivation to comply with the wishes of these individuals. To come up with the three items that measured subjective norm, the normative beliefs were weighted by the corresponding motivation to comply.

Table 6: Decision Rules – Formative versus Reflective[92]

	Formative Model	Reflective Model
1. Direction of Causality		
• How is the direction of causality from construct to measure implied by the conceptual definition?	Items → Construct	Construct → Items
• Are the items defining characteristics or manifestations of the construct?	Characteristics	Manifestations
• Would changes in the items cause changes in the construct?	Yes	No
• Would changes in the construct cause changes in the items?	No	Yes
2. Interchangeability		
• Are items interchangeable?	Not necessarily	Yes
• Should the items have the same or similar content? Do the items share a common theme?	No	Yes
• Would dropping one of the items alter the conceptual domain of the construct?	Yes	No
3. Covariation		
• Is covariation among the items expected?	Not necessarily	Yes
• Should a change in one of the items be associated with changes in the other items?	Not necessarily	Yes
4. Nomological Net		
• Should the nomological net of the construct items differ?	Yes	No
• Are the items expected to have the same antecedents and consequences?	No	Yes

Having considered the decision rules of Jarvis et al. (2003) (see Table 6), I decided to measure subjective norm formatively instead of reflectively, although this is in contrast to Bock et al. (2005) and Jeon et al. (2011).[93] I based my decision on the following: Firstly, subjective norm is composed of the perceived social pressure caused by the CEO, the immediate supervisor, and colleagues because these three groups of people are expected to represent all of the possible sources for social pressure within a professional context (cf. for example Karahanna et al. 1999, p. 201). Following this argument, the items are considered to define the construct. The causality is directed from the items to subjective norm. Secondly, if the social pressure caused by the colleagues was replaced (e.g., by the social pressure caused by former supervisors or family members), it could be assumed the subjective norm construct would change, i.e., the change of an item is expected to alter the construct. Conversely, modifications in subjective norm are not necessarily associated with a change in items. Thirdly, if one of the three groups of people were omitted and only the items related to the other two parties were used to measure subjective norm, the conceptual domain of the construct would be modified, i.e., the exclusion of one or more items would change the conceptual domain of subjective norm. Fourthly, if for example the value of the

---

[92] Jarvis et al. 2003, p. 203
[93] According to Jarvis et al. 2003 and Fassott 2006, the majority of constructs are correctly specified as reflective or formative. Nevertheless, both articles say a considerable number of studies incorporate incorrectly modeled constructs. When a misspecification was detected, constructs were mainly modeled as reflective even though formative measurement would have been more appropriate.

item relating to colleagues increase due to a stronger normative belief and/or an increase in the motivation to comply with colleagues' interests, it does not necessarily have an effect on the items relating to the CEO or the immediate supervisor. Therefore, it is not expected that the value change of one item will immediately lead to changes in the value of other items. The last reason for measuring subjective norm formatively refers to antecedents and consequences of items. Colleagues, for example, might create social pressure because of fellow employees' desire for affiliation (see Murray 1938; Hill 1987). The immediate supervisor, on the other hand, will tend to create social pressure by way of a position of power (cf. Blau 1964, pp. 115ff.; Molm 1997, pp. 29ff.). Thus, the antecedents of items are considered to be different. Furthermore, the perceived social pressure caused by the immediate supervisor is likely to have consequences other than the social pressure generated by colleagues, i.e., the consequences of items are expected to be diverse. Besides the arguments based on the rules of Jarvis et al., the formative measurement of subjective norm is further supported by other researchers (cf. for example Eckhardt et al. 2009, p. 15; Hsieh et al. 2008, p. 106; Herath and Rao 2009, p. 115; Karahanna et al. 1999, pp. 196f.; Limayem et al. 2000, pp. 424f.; Plant 2009, pp. 187ff.).[94]

4.4.1.3 Perceived Behavioral Control

For the operationalization of perceived behavioral control, which reflects the intuitive opportunities and capabilities for exchanging knowledge with external partners in OI-projects, I again followed the instructions of Ajzen (2002a). I considered direct measures preferable (cf. Ajzen 2002a, pp. 6f.). Due to the evolved differentiation between perceived self-efficacy and perceived controllability (see Ajzen 2002b and chapter 3.1.3 for more details on the differentiation), I decided to operationalize perceived behavioral control by using items intended to measure self-efficacy and items anticipated to measure controllability. Even though several researchers have dealt with this differentiation of perceived behavioral control in their quantitative studies (see Terry and O'Leary 1995; Sparks et al. 1997; Manstead and Eekelen 1998), only Armitage and Conner (1999a; 1999b) gave an overview of all the items they had used to measure self-efficacy and controllability. They additionally applied a principal component analysis to categorize the items as either self-efficacy or controllability. Therefore, I based the operationalization on the seven items listed by Armitage and Conner (1999a). For reasons of linguistic consistency within the questionnaire, I reformulated items with question design into statements.[95] Having done so,

---

[94] The discussion points out that opinions regarding the correct specification of subjective norm (i.e., formative versus reflective) differ. Out of curiosity, I calculated two versions of the theoretical model: Firstly, I calculated the model with a formatively measured subjective norm (mode B; cf. Becker et al. 2012, p. 365). Secondly, I computed the same model, but with a reflective measurement of subjective norm (mode A). The results of the two calculations showed no meaningful qualitative difference.

[95] The item *"To what extent do you see yourself as capable of eating a low-fat diet?"* (Armitage and Conner 1999b, p. 41) might be redrafted as: "I am capable of eating a low-fat diet".

two of the reformulated items became very similar to items that were originally in the statement design. To avoid redundancy, the two reformulated items (one self-efficacy item, one controllability item) were deleted without replacement. Consequently, five items were left.[96] The wording of these remaining items was adapted to the OI-context and a 5-point Likert scale ranging from "strongly disagree" to "strongly agree" was applied. In addition to the five items suggested by Armitage and Conner (1999a), a sixth item regarding behavior's difficulty to perform was employed, which has often been applied in past research to measure self-efficacy (see e.g., Terry and O'Leary 1995; Sparks et al. 1997; Manstead and Eekelen 1998). The phrasing of the item was adapted to the OI-context and measured on a 5-point Likert scale ranging from "very difficult" to "very easy". In the end, perceived behavioral control was operationalized using six items, which measured the construct reflectively.

4.4.1.4 Intention

For the operationalization of R&D employees' intention to exchange knowledge with external partners in OI-projects, I again followed the instructions of Ajzen (2002a) and applied direct measures (cf. Ajzen 2002a, p. 4). On the basis of Bock et al.'s (2005) scale for measuring the intention of Korean employees to share knowledge – which has also been applied in a slightly modified version by Chow and Chan (2008) and Jeon et al. (2011) – the wording of the items was adapted to the context of knowledge exchange in OI-projects. According to Bock et al. (2005) and for the purpose of distinguishing between the intention to exchange documented (explicit) knowledge and undocumented (implicit or tacit) knowledge with external partners in OI-projects, intention was operationalized as a Type I reflective second-order construct[97] (reflective first-order constructs, reflective second-order construct; both measured in mode A) using a total of five items.[98] All items were measured on a 5-point Likert scale ranging from "very unlikely" to "very likely". Two of these were expected to measure the intention to exchange documented knowledge and three items were supposed to account for the intention to exchange undocumented knowledge (see Table 4). Consequently, the lower order components did not have exactly the same number of items. However, the imbalance between the measures of intention to exchange documented and undocumented knowledge was considered passable. Therefore, the repeated indicator[99]

---

[96] According to the principal component analysis carried out by Armitage and Conner 1999a, three of the items would measure self-efficacy and two would measure controllability.
[97] In principle, a distinction is drawn between one-dimensional and multi-dimensional constructs. One-dimensional constructs consist of only one component. Multi-dimensional constructs (e.g., second-order constructs) would be expected to have multiple dimensions. For more details on multi-dimensional constructs or second-order constructs see Becker et al. 2012; Edwards 2001; Jarvis et al. 2003; MacKenzie et al. 2011; Ringle et al. 2012, pp. appendix B; Wetzels et al. 2009
[98] Chin 1998a, p. x claimed that second-order constructs must be related to other constructs in the research model so they are the consequence or predictors of other latent variables. This study fulfilled this criterion.
[99] Indicator and item are equivalent terms and used interchangeably.

approach (also known as indicator reuses approach), which is recommended for Type I second-order constructs (cf. Lohmöller 1989, pp. 130ff.; Wold 1982, pp. 40ff), was applied to analyze the second-order model – even though the approach actually works best with an equal number of items (cf. Becker et al. 2012, pp. 365f.; Ringle et al. 2012, pp. appendix b).

4.4.1.5 Enjoyment in Helping

The operationalization of enjoyment in helping was based on the measures of Kankanhalli et al. (2005), which has also been used by Lin (2007b) and Jeon et al. (2011). The wording of the four items was adapted to the context of knowledge exchange in OI-projects. During the main pre-test (see chapter 4.4.2), half of the respondents stated that two of the four items were very similar and, thus, hard to distinguish. The respondents were asked which of the two questions they considered more comprehensible and meaningful. The item that was assessed to be more difficult to understand was excluded from the final questionnaire. The remaining three items were reflectively measured on a 5-point Likert scale ranging from "strongly disagree" to "strongly agree".

4.4.1.6 Sense of Self-Worth

Sense of self-worth was measured using five items adapted from Bock et al. (2005). These items have also been employed by Huang et al. (2008), Teh and Yong (2011), and Tohidinia and Mosakhani (2010), even though Tohidinia and Mosakhani named the construct differently. The wording of the five items was adapted to the OI-context. The items were reflectively measured on a 5-point Likert scale ranging from "strongly disagree" to "strongly agree".

4.4.1.7 Reciprocity

The operationalization of reciprocity was based on the measurement by Kankanhalli et al. (2005), which has also been applied by Jeon et al. (2011). The items' wording was adapted to the OI-context and reflectively measured on a 5-point Likert scale ranging from "strongly disagree" to "strongly agree".

4.4.1.8 Rewards

For the operationalization of rewards, I followed the measurement by Kankanhalli et al. (2005), which has been used in a slightly modified version by Lin (2007b) and Tohidinia and Mosakhani (2010). The wording of the five items was adapted to the OI-context. During the main pre-test (see chapter 4.4.2), three of the four respondents indicated (independent from each other) the literature-based incentives given in the questionnaire would not motivate R&D employees to exchange knowledge with external partners in OI-projects. In fact, they would be more motivated by things not directly granted from within the organization. When they were asked about factors that would better motivate R&D employees to exchange their knowledge in OI-projects, all pre-testers stated similar things. These were gathered and the wording was adapted to the other reward items. As a result, five additional items were added to the original five reward items from Kankanhalli et al. (2005). In the end, reward was measured using ten items (see Table 5), which were all measured reflectively on a 5-point Likert scale ranging from "strongly disagree" to "strongly agree".

4.4.1.9 Control Variables

To control for aspects that could influence the results of the theoretical model without being explicitly part of it, information about employees' OI-experience, employees' personality, demographic data and work-related characteristics were collected.

The first control variable relates to the experience with different OI-partners. Employees were asked how often they had collaborated on OI-projects with universities/research institutes, customers, suppliers, competitors, and other industrial partners. The employed 5-point Likert scale ranged from "very rarely" to "very often".

The second control variable relates to the number of OI-projects. Employees were asked to state the number of OI-projects during the last three and the last ten years, respectively.

The third control variable relates to the personality of employees. By employing the ten-item personality inventory used by Gosling et al. (2003), the "Big Five" main personality traits (extraversion, agreeableness, conscientiousness, emotional stability, and openness to experience) were measured. Each personality trait was measured using two items, one of which was a reversed coded item. All items were measured on a 5-point Likert scale ranging from "strongly disagree" to "strongly agree".

To control for the influence of demographics, employees were asked to state their age, gender (male/female), highest educational degree (drop-down menu: apprenticeship, bachelor degree, master degree/diploma, PhD degree/doctorate, others) and to name their field of education (e.g., chemical physics, material science, business administration).

The last control variables were work-related. Employees were asked to state their tenure and the country where they work.[100]

In addition to the control variables, an open-ended question regarding important requirements for knowledge exchange in OI-projects was integrated into the questionnaire. Employees were requested to state up to five requirements that would allow them exchange their knowledge with external partners in OI-projects. Respondents could also give feedback at the end of the survey.

---

[100] Two of the four participating companies do not have a combined department for R&D; instead they differentiate between research and development. Since both companies asked for a differentiated analysis of their company-specific results, an additional question was added to their questionnaire and employees were asked about their functional affiliation.

4.4.2 Pre-test

The pre-test was conducted in March and April, 2012, and consisted of two steps. Firstly, four scholars assessed the suitability of scales and the comprehensibility of the original, i.e., English questions and possible answers. Their feedback was compiled and the feasibility of their suggested adaptations was assessed. If their suggestions improved comprehension without altering the original item or its construct significantly, the changes were adopted. Following feedback from the first pre-testers, the main pre-test with four representatives of the participating companies was conducted.[101] Three pre-testers checked the German survey, one the English version. All pre-testers received an e-mail with the link to the online survey (where they could select the language) and a print version of the questionnaire. The pre-testers were requested to complete the questionnaire online and note their comments on the printed version, so that we could discuss their feedback during a 30-minutes' debriefing call. The pre-testers were asked to pay special attention to the comprehensibility of the questions and possible answers; the fit between questions and scale; and the structure of the questionnaire. Furthermore, they were asked to keep track of the time they would need to complete the survey. After all debriefing calls were conducted, the feedback was again compiled and the feasibility of suggested modifications was assessed. The pre-test involving the company representatives led to some minor changes in wording and two major changes in the questionnaire: One of the items designed to measure enjoyment in helping was excluded due to perceived redundancy (see chapter 4.4.1.5). The pre-test feedback also suggested it would be worthwhile to complement the existing reward measures with five additional items to either improve the measurement of the existing reward construct or to introduce another type of reward that might be expected to have a stronger impact on R&D employees' attitude.[102] In case that the later conducted exploratory factor analysis (see chapter 6.4.1.1) suggested to differentiate between two separate reward constructs, the hypothesis regarding rewards derived in chapter 3.3.5 would need to be sub-divided as follows, where "reward A" would be measured using the original items derived from the literature and "reward B" using the pre-test items:

> H7a: Reward A has a negative impact on R&D employees' attitude toward exchanging their knowledge with external partners in OI-projects.
>
> H7b: Reward B has a positive impact on R&D employees' attitude toward exchanging their knowledge with external partners in OI-projects.

---

[101] The pre-testers had been assigned by the direct contact in the respective companies.
[102] As already explained in chapter 4.4.1.8, the pre-testers indicated the literature-based incentives detailed in the questionnaire would not motivate them. Instead, they suggested five other factors with a higher potential to motivate R&D employees to exchange knowledge with external partners in OI-projects. In contrast to the five original reward items – which drew mainly on hygiene factors – the five factors added by the pre-testers are associated with motivators (see Herzberg et al. 1959; Herzberg 1968).

4.4.3 Data Collection

Each of the four participating companies prepared a mailing list. All of the employees on the list were expected to have experience with OI-projects, during which they had personal contact with external partners. Company A appointed 93 relevant people, Company B named 40 R&D employees, Company C prepared a list with 135 names, and Company D appointed 15 personnel (see Table 7).

To contact the identified employees and distribute the link, I developed a cover letter for each company. All cover letters had the same structure and comprised a short introduction detailing information on myself and the research topic, an emphasis on the importance of participation for the company and the research project, a deadline and the request to complete the survey within the given time (about 14 days), a link to the online survey[103], and advice on how to choose the appropriate language (as the questionnaire would be available in German and English). The employees were also informed the survey was anonymous, responses could not be linked to personal data, and it would take 7-10 minutes to complete. Employees were encouraged to send me an e-mail if they had any questions.

Companies A and C decided to independently distribute the survey link to the pre-defined group of employees. The other two companies personally informed each assigned employee about the survey and requested their participation. However, I distributed the link after I had received the mailing list. The data collection began in the middle of June 2012, when the survey link was distributed to employees of companies A and B. It was completed in the middle of August 2012; one month after the last employees had received the survey link. As detailed in the cover letter, employees had 14 days to complete the questionnaire. Since the data collection period overlapped with German summer holidays, they were granted an additional week before they received a friendly reminder.

4.4.4 Data Cleansing, Data Preparation and Final Sample

In sum, 283 employees were contacted but only a total of 199 clicked on the survey link. 46 of the 199 viewed the welcome page of the survey but did not then proceed to the first question. These cases were immediately excluded from the sample (see Figure 14). The remaining 153 cases were investigated with respect to missing data and outliers: 21 questionnaires had not been completed so special attention had to be paid to the missing data and the application of remedies in these cases. The other 132 respondents had finished

---

[103] Each company had its own link to allow responses to be assigned to the companies. However, the questionnaires were identical except for the single question regarding an employee's functional affiliation.

the survey and answered all of the mandatory questions (see chapter 4.4.1).[104] Nevertheless, missing data in terms of invalid values and omitted information on age and/or gender as well as outliers was also discovered. 20 of the 153 responses were finally excluded from the sample.

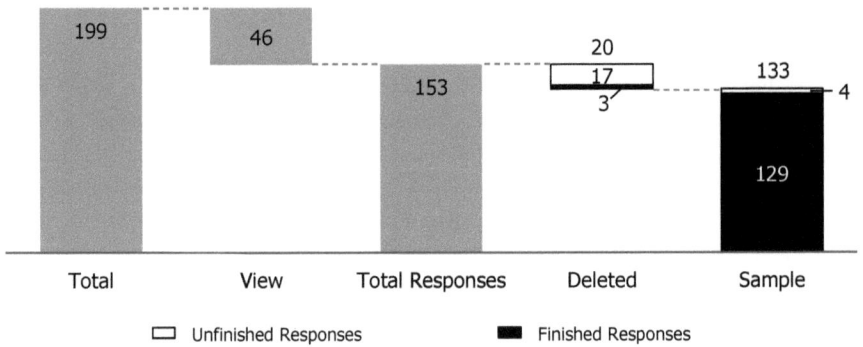

Figure 14: Overview Data Cleansing[105]

4.4.4.1 Missing Data

In general, missing data can only be ignored if they are part of the research design. Since this is not the case in this study, the extent of missing data (per response/case and per item) and the randomness with which the missing data occur had to be determined so as to decide an appropriate missing data remedy (cf. Hair et al. 2008, pp. 44ff.).[106] According to Little and Rubin (1989), there are three strategies for handling missing data: imputation (i.e., missing values are replaced by suitable approximations); weighting (i.e., incomplete cases are excluded from the analysis and the remaining complete cases given a new weight); and analyzing incomplete data directly. The most appropriate strategy depends on many factors, including the extent and randomness of missing data (see Graham et al. 2012; Hair et al. 2008). Each of the strategies has its advantages and pitfalls.[107] Therefore, a certain bias has to be accepted in cases of missing data.

---

[104] Eight of the 132 respondents did not state their gender and/or age because these details were not mandatory and could be ignored without canceling the survey.
[105] Author's illustration
[106] According to Hair et al. 2008, p. 47, the extent of missing data is determined by the percentage of items with missing values per case and the percentage of cases with missing values per item. According to Cole 2008, pp. 216f., three types of randomness can be distinguished: Data can be missing completely at random (MCAR); missing at random (MAR); or missing not at random (MNAR). In this study, all missing values were considered to be completely at random (MCAR).
[107] For a more detailed discussion see Allison 2001; Graham et al. 2012; Schafer et al. 2002

As mentioned in chapter 4.4.4 and shown in Figure 14, 21 of 153 respondents did not finish the survey. The share of missing values differed considerably by case and ranged from 2.9% to 95.6%. To enable a meaningful data analysis, I followed the recommendation of Hair et al. (2008, p. 48) and conducted listwise deletion (also known as casewise deletion or complete-case analysis, cf. Graham et al. 2012, p. 155) in 14 cases, where missing data affected the dependent variable "intention". A further two cases were disqualified due to employee feedback where some of the questions had been answered "neutral" for personal reasons.[108] Another case with a 32.4% share of missing values and a complete randomness in missing data (MCAR) was deleted because the missing data affected all five items of the "perceived behavioral control" construct. The remaining four (out of 21 unfinished) responses were usable for the analysis, even though they were not completely filled out. The share of missing values per case ranged from 14.7% to 23.5%, but the missing data only concerned the control variables and occurred completely at random (MCAR).

132 of 153 employees finished the survey. Due to the feedback of three employees who were unable to give meaningful answers, these were also removed from the sample. Eight of the remaining 129 respondents had only answered all of the mandatory questions and so did not state their gender and/or age. In addition, one of these eight respondents obviously did not reply reasonable to the personality-related questions (Big Five), i.e., these Big Five related values were declared missing. The declaration of missing data due to invalidity was also made in three other cases: The input regarding the field of education was not usable twice and one case showed equal values for age and tenure, which suggests an entry error. In summary, 129 of the completed responses were considered usable for the analysis – of these, 118 were entirely complete. Only one of the 11 cases with missing data had a share of missing values above 10%.[109] Furthermore, missing data only concerned the control variables and occurred completely at random (MCAR).

Overall, after identifying all of the responses containing missing data and conducting listwise deletion for all of the cases where this remedy was considered most appropriate, a total of 133 responses was left. 15 of these 133 responses still had missing data but only control variables were affected in these cases. The share of missing value per case ranged from 1.5% to 23.5%. Based on the 133 cases, the share of missing value per control variable ranged from 2.3% to 8.3%. Since the missing data did not affect model-relevant items and the share of missing value was mainly beyond 20%, a pairwise deletion (also known as available-case analysis, cf. Graham et al. 2012, p. 155) was considered more applicable for the 15 cases than replacing the missing values with calculated values (cf. Cole 2008,

---

[108] It was not possible to say which questions had been answered "neutral" and which had been answered truthfully using the "neither... nor..." option, i.e., the missing values could not be clearly identified. Therefore, both cases were excluded from analysis.
[109] According to a rule of thumb of Hair et al. 2008, p. 47, missing data can be ignored if the share of missing value per case is under 10%.

pp. 271ff.; Hair et al. 2008, pp. 44ff.).[110] Therefore, pairwise deletion was conducted during the data analysis of these 15 cases.

4.4.4.2 Outliers

To identify outliers, a non-recursive procedure formulated by Selst and Jolicoeur (1994) was employed. If sample size is at least 100, this approach suggests a cut-off score of 2.5 standard derivations from the mean. By applying this threshold, three outliers were detected within the sample of 133. All outliers were related to the control variable that evaluates the OI-experience by requesting the number of OI-projects during the last three and ten years respectively. Since the identified outliers concerned only one control variable and all other answers seemed reasonable, the outliers were not excluded from the analysis. Instead, truncation was applied, i.e., the conspicuous values relating to the number of OI-projects were replaced by the next possible plausible value, which maintained the order of data and simultaneously abated distributional issues (cf. Osborne 2008b, p. 208).

After considering missing data and outliers, the data cleansing process led to a final sample size of N = 133 (118 cases were complete and 15 responses had missing values at control variable related items). Table 7 gives an overview of the sample.

Table 7: Sample, Firms, and Responses

	Contacted Employees	View Only	Unfinished Responses	Finished Responses	Usable Responses	Response Rate[111]
Company A	93	19	7	56	58	62.4 %
Company B	40	2	2	36	33	82.5 %
Company C	135	23	5	34	35	25.9 %
Company D	15	2	7	6	7	46.7 %
Total	283	46	21	132	133	47.0 %

4.4.5 Method of Data Analysis

In this study, more than three variables[112] had to be analyzed simultaneously. Therefore, only multivariate data analysis methods could be employed (cf. Bryman 2008, p. 330). Of all multivariate methods (e.g., regression, conjoint analysis, cluster analysis),[113] structural equation modeling in particular has attracted the attention of social science researchers and interest in this application has been growing (cf. Anderson and Gerbing 1988, p. 411; Chin 1998a, p. 7; Rigdon 1998, p. 251; Shook et al. 2004, pp. 397ff.). This is mainly due to its

---

[110] As already mentioned, every strategy for handling missing data, i.e., listwise deletion, pairwise deletion and (any kind of) imputation all have inherent bias (see Cole 2008). Therefore, the decision comes down to a choice of which bias is the most acceptable.
[111] The response rate equals the usable responses divided by contacted employees.
[112] In the following, variable and construct are equivalent terms and used interchangeably.
[113] For an overview of methods for multivariate data analysis see Hair et al. 2008; Backhaus et al. 2011.

ability to analyze multiple relationships between numerous independent and dependent variables simultaneously (see Gefen et al. 2000). This enables researchers to develop causal models and to test related hypotheses (see Bagozzi 1980). Another decisive benefit is that latent, unobservable constructs can be included and considered in the analysis (see Fornell and Larcker 1981).

Structural equation models consist of a measurement model and a structural model (see Backhaus et al. 2011; Hair et al. 2008; Jarvis et al. 2003; Weiber and Mühlhaus 2010). The measurement model, which is also known as the outer model comprises a selection of manifest items that are directly measurable or observable and related to a certain construct (see Fornell and Larcker 1981; Tenenhaus et al. 2005). The structural model, which is also known as the inner model, represents the assumed causality between the constructs (see Anderson and Gerbing 1988; Tenenhaus et al. 2005).

Structural equation models can be differentiated between covariance-based and variance-based approaches (see Backhaus et al. 2011; Dijkstra 1983; Fornell and Bookstein 1982; Fornell 1987; Gefen et al. 2000; Lohmöller 1989; Tabachnick and Fidell 2007). The covariance approach can be traced back to the work of Jöreskog (1970; 1973). The estimates of the parameters are based on the covariance matrix and maximum-likelihood is applied. The covariance approach is implemented in software applications such as LISREL (linear structural relationships, see Jöreskog and Sörbom 2001) and AMOS (analysis of moment structures, see Arbuckle 2006). The variance-based approach, on the other hand, is attributable to Wold (1966; 1975; 1982). This is a component-based approach, which uses a partial least squares procedure (PLS) to estimate the model. This approach is implemented in software applications such as SmartPLS and PLS-Graph (see Temme et al. 2010).

Since covariance-based and variance-based estimating approaches are based on different analytical techniques, the application areas and the demand on data differ. In order to select the most appropriate approach, I followed the decision-making criteria suggested by several researchers (cf. Chin and Newsted 1999, p. 337; Hair et al. 2011, pp. 143f.; Hair et al. 2012, pp. 419ff.; Henseler et al. 2012, pp. 261ff.; Weiber and Mühlhaus 2010, pp. 65ff.). These criteria mainly refer to statistical properties (e.g., latent variable score), data characteristics (e.g., sample size, distribution), model characteristics (e.g., model complexity, formative measures), and research objectives (e.g., theory testing, prediction) (cf. Henseler et al. 2012, pp. 261ff.). In this study, the following four determining criteria mainly influenced the decision to use PLS, i.e., variance-based structural equation models:

- ✓ Applicability for small sample sizes – In contrast to covariance-based structural equation models, PLS is able to provide robust outcomes even if the number of observations is low (cf. Chin and Newsted 1999, pp. 326ff.; Ringle et al. 2009, pp. 22ff.). Chin (1998b, p. 311) introduced a rule of thumb for PLS, saying that the minimum requirement regarding sample size equates to ten times the highest number of items pointing to a formatively measured construct or ten times the highest number of structural paths pointing to a latent construct.[114] Relating to my study, where the dependent variable "attitude" features the most complex formative relationship[115], this rule of thumb suggested a minimum of 50 observations. Thereby, the final sample size of N = 133 (see chapter 4.4.4) exceeded the minimum requirement by far. However, the 133 observations would not be sufficient for the covariance-based approach (cf. Chin and Newsted 1999, pp. 309f.; Sosik et al. 2009, p. 16), which suggested the usage of PLS.

- ✓ Convenient integration of formative constructs – Generally, it is possible to integrate formative measures in PLS and in covariance-based structural equation models (see Bollen and Davis 2009; Diamantopoulos and Riefler 2011). However, to guarantee model identification in covariance-based approaches, researchers must consider specific constraints regarding the model, which often contradict the theoretical framework (cf. Hair et al. 2012, p. 420). This is contrary to variance-based structural equation models, where the involvement of formative measures tends to be relatively unproblematic (cf. Henseler et al. 2009, p. 290; Henseler et al. 2012, p. 267). Therefore, PLS is the recommended alternative to covariance-based approaches if formative measurement models are incorporated into the theoretical framework (cf. Diamantopoulos and Winklhofer 2001, p. 274). Since the subjective norm construct was measured formatively in this study (see chapter 4.4.1.2), PLS was considered most suitable.
- ✓ No distributional assumptions – Contrary to covariance-based structural equation models, PLS does not require the normal distribution of data for a stable parameter estimate (cf. Reinartz et al. 2009, p. 336), i.e., non-normality does not cause estimation bias in PLS (cf. Hair et al. 2011, pp. 143ff.). Even though Reinartz et al. (2009) demonstrated that non-normality does not necessarily have a negative effect on the quality of covariance-based estimates, Ringle et al. (2009) showed that the combination of non-normality and formative measures in covariance-based structural equation models lead to significant losses of accuracy and robustness. Since the majority of data in this study was not normally distributed (see chapter 6.1.1) and a formative construct was incorporated, it was advisable to apply the variance-based approach (cf. Hair et al. 2011, p. 144; Ringle et al. 2009, pp. 17ff.).
- ✓ Focus on prediction rather than theory testing – Most scholars agree covariance-based approaches are the most suitable for testing a theory, whereas variance-based structural equation models are most appropriate for predicting relationships and (further)

---

[114] The minimum requirement for a sample size in PLS is defined by the larger number of the two multiplications.
[115] Five structural paths (enjoyment in helping, sense of self-worth, reciprocity, reward A, and reward B) point to the attitude construct.

developing theories (see e.g., Chin and Newsted 1999; Hair et al. 2012; Lehner and Haas 2010, p. 81). This is particularly reasoned in the requirement of covariance-based approaches to model all theoretical relationships in order to test a theory as a whole, while PLS is not that strict and, therefore, can be used to test portions of a theory. (cf. Sosik et al. 2009, pp. 12ff.). The TPB, which contributed considerably to the research model of this study, is a well-established theory (see chapter 3.1). The research objective was not to test the entire theory; rather the goal was to predict employees' intention to exchange knowledge in OI-projects as well as their attitude toward knowledge exchange in OI-projects and to explain as much of the variance as possible of these two dependent variables. Furthermore, the study tried to explore the factor most likely to influence attitude and intention. Therefore, the application of PLS was recommended (cf. Hair et al. 2011, pp. 143f.).

As explained above, variance-based structural equation modeling was considered the most appropriate approach and, therefore, PLS was chosen as the data analysis method. The analysis was conducted by applying SmartPLS 2.0 (M3) Beta and PASW Statistics 18.

4.5 Chapter Summary

In consideration of the advantages and disadvantages of qualitative and quantitative research methods, a mixed-method approach using three phases was adopted. Firstly, 12 R&D managers were interviewed and MaxQDA was used for the analysis. Secondly, an online survey among R&D employees was conducted to test the research model and related hypotheses. The research model was operationalized based on existing constructs and items. The pre-test added five items (related to the reward construct) to the final questionnaire. After the data cleansing, a final sample of 133 usable responses remained. The data were analyzed through variance-based structural equation modeling (PLS). Lastly, the findings were discussed with scholars and R&D representatives from two participating companies during three follow-up group discussions.

5 Findings from Qualitative Pre-Study (Interviews)

This chapter summarizes the findings from interviews conducted with R&D managers.[116] The first sub-chapter provides insights into their understanding of open innovation. In the second sub-chapter, the typical procedure for setting up an OI-project is described. Thereafter, the focus lies on the search and selection by companies of an appropriate OI-partner. The fourth sub-chapter deals with the basic conditions for an OI-project. Finally, the advantages and challenges of open innovation are considered.

5.1 Open Innovation from an R&D Perspective

Since the participating companies in this study had publicly communicated the importance of an OI-approach, I expected open innovation would be a very relevant topic within their R&D departments. The interviews confirmed this expectation: Open innovation was repeatedly highlighted as a valuable and irreplaceable pillar of R&D in all of the companies. In the most general sense, open innovation reflects for the interviewees their company's ambition to open up the innovation process and to co-operate with external partners. Relating this interpretation to the project portfolio of R&D departments means that every innovation-related activity, project, or alliance conducted with at least one external partner can be categorized as open innovation. Consequently, strategic alliances with the purpose of collective innovation are a form of open innovation:

> "Well, I assume that [...] I have at least a very broad definition of open innovation. Basically, it involves all innovation plans, all research projects, research co-operation that we undertake with someone who is external to the company and possesses special competencies." (A4)

But besides broadly formulated definitions, interviewees also stated concrete characteristics of open innovation. Two aspects were especially emphasized and repeatedly highlighted as important elements. The first aspect refers to a company's own state of knowledge and resulting demands on the skills of the external partner:

> "Open innovation means that we look for external partners [...] beyond the internal scientific [and] technological state of knowledge in order to accelerate the development process for new products." (B2; addendum by author)

Hence, a central element of open innovation is that the external partner brings in expertise the host company does not have – or has only to a limited degree. This external know-how

---

[116] The interviews were conducted in German. Therefore, all personal quotations in this chapter were translated into English faithful to the original. The findings were partially published in Herstatt and Nedon (2014).

complements the internal knowledge base and so contributes to the improvement of a company's innovation process:

> "The point is: Which external partners – possessing expertise that we do not have – can be connected reasonably and cleverly? How can I create value for the own company out of this? So basically, how can I establish a scientific network and extract things out of this scientific network that helps me to either start up or accelerate my innovation process?" (A4; anonymized by author)

> "[We are looking] for partners, who possess certain technologies we do not have in-house, but where we have the feeling or the opinion that the combination of our technology with their technologies could offer an innovation advantage." (B2; addendum by author)

A second aspect repeatedly emphasized was the mode of co-operation. The interviewees regard open innovation as close, intensive, and systematic co-operation with a partner, i.e., problems are brought to external partners with a view to finding solutions. Open innovation means becoming involved in a permanent and iterative exchange with a partner. Through intensive collaboration with different partners, a dynamic network with a small static core evolves. This is not only desirable – it is intentionally fostered in R&D:

> "[Open innovation does not just mean to] place assignments externally, let the project be executed, and ask after two years: 'What's the solution?' It is rather a permanent, iterative exchange and the allocation of employees from the research department [...] who are able to competently accompany it." (A5; addendum by author)

Another finding from the interviews was that the R&D managers distinguish between the forms of open innovation described above (i.e., close collaboration with iterative exchange) and open, anonymous, worldwide calls for proposals (i.e., crowdsourcing). They tend to be skeptical of the second approach. Although they consider crowdsourcing as a form of open innovation, they rarely apply it – especially in comparison with other OI-approaches. As an explanation for this imbalance, interviewees referred to some challenges associated with crowdsourcing. Firstly, it can be difficult to explain a problem without the need for follow-up discussions. R&D managers also view the lack of confidentiality inherent in crowdsourcing as a deterrent. Thirdly, an open request for a proposal is a black box, i.e., the output of such an open call can hardly be controlled. Lastly, crowdsourcing is perceived as a competition to internal efforts.

In summary, the R&D managers generally defined open innovation as the extension of the innovation process and the integration of external partners. However, from their perspective two additional elements characterize open innovation and complement the definition. Firstly, it reflects the combination of internal and external knowledge and the pooling of mostly complementary expertise. Secondly, open innovation stands for the close co-operation with at least one external partner, in which the partners learn from each other through intensive knowledge exchange and so advance innovation processes. Another finding was that the exchange of knowledge, which can be interpreted as the coupled OI-process, is in the forefront at their companies:

> "If we start co-operations with academic partners, we bring knowledge to the outside to such a degree as we name questions openly. [...] And alone through the definition of this issue and the elaboration for the colleagues at the university, it is not just obtaining things they are doing, but rather we [...] engage in advance by telling things and describing problems and expect nothing till the point that we get something back." (A6)

5.2 Setting up OI-Projects

During the product development, ideas normally pass through a stage-gate process (see Cooper 1990, 1996) – regardless of whether external partners are involved or not. However, if a company wants to benefit from open innovation, additional steps are necessary:

> "Since one year, we have this face-gate process, where the innovation actually starts with me considering [...] whether it fits with [our] corporate strategy. What are the underlying business potentials? Is [...] an above-average growth possible? We always have this process. Whether I do this closed loop or with a partner [...] I think the process is no different. Of course, open innovation is a little more complex, because you have a higher need for co-ordination." (A1, anonymized by author)

When asked about any additional steps intrinsic to the OI-process, interviewees give very similar responses, which tend to coincide with the literature (for an overview see chapter 2.1.3.3.). Initial tasks in particular are described in detail, indicating the effort that tends to be associated with this phase. According to interviewees' answers, the innovation process starts by identifying strategically important fields of innovation that have to be developed. Once the target is defined, the first step toward open innovation is to examine whether all of the required competencies are internally available or have to be found externally. If external expertise is required, brainstorming sessions to identify suitable partners then take place. A list of favored potential partners is compiled and these are contacted via telephone or e-mail initially on a non-binding basis. If this initial contact leads to a mutual consensus, the parties work toward a non-disclosure agreement (NDA). Co-operation only begins after this – and possibly other supplementary documentation – has been signed:

> "We specify the problem and define it for ourselves; define where we have fields that we cannot answer or where we cannot solely contribute to the problem solving or contribute not too well. [We] then make a list of potential external experts – who can help us much better than we could do internally – and approach them purposefully and inquire to what extent there is, firstly, willingness and, secondly, also the possibility to help us [...]. That way, we then try to get the necessary expertise quasi "in-house" through research projects, joint research projects. [...] We search for the one or two or three or four – as the case may be – luminaries for a concrete problem that we have and approach them

very purposefully and usually conclude a confidentiality agreement, a joint development agreement, i.e., everything very stipulated by contract, and then enter the project with them." (A4; addendum by author)

After all parties have signed an NDA, the next important step is the allocation of IP rights. As soon as this task is successfully accomplished, the OI-project set-up is written down in a project plan with milestones and responsibility assignment. This project plan is particularly important for project control, which among other things tracks if all partners contribute according to their assigned responsibilities:

"It starts with [the challenge] that one has to first agree on how the IP rights will be allocated. This is always the most complex and hardest and above all also the most important in the open innovation business; that this is resolved before one begins. If one has achieved this, then one has to draw up a plan together [with the partner]. Once this is done, one has to ensure that each partner really delivers what he had promised."
(A3; addendum by author)

As already indicated by the definition of open innovation as close co-operation and intensive knowledge exchange with external partners, several interactive mechanisms appear throughout the whole OI-process. Interviewees provided further insights into the intensity and frequency of interactions:

"One has regular [...] meetings. They can be in a monthly, bi-monthly, four-month interval. Research is sometimes not so fast. And then the project status is defined; next steps; who is in charge of what." (B2)

"In principle, the results must be discussed later on, of course. [...] That means a constant exchange actually always takes place. And I think, how intense or how strong this exchange is, always depends a bit on how the project itself is organized. This does not depend so much on the partner, but rather it is a question of the project and the target, which I associate with such a project." (B3)

Usually, these interactions and the entire OI-project do not end before the project goals are achieved and there is a solution to the initial problem. However, finishing one OI-project is often just the beginning of a new OI-cycle:

"Based on our knowledge, we globally pick the experts, who can help solving a problem [...] and work with them under strict confidentiality until the problem is solved. And then a new problem arises and I choose somebody else, who [...] can help me best with this problem. This is quasi the ideal case, the silver bullet." (A4)

To sum up, an OI-process basically does not significantly differ from a closed innovation process. However, open innovation involves some extra steps associated with additional effort. The course of action during OI-projects described by the interviewees is consistent with the literature. The interaction mechanisms stated in the literature were given particular mention by several R&D managers.

5.3 Searching and Choosing OI-Partners

Not every OI-project the surveyed companies are engaged in involves a pro-active search for partners. Some OI-projects are the outcome of self-initiative by partners; others result from regular meetings with customers and suppliers – especially in the B2B-context. However, if a suitable partner is still required companies often – as already noted – search globally and interdisciplinary for complementary expertise:

> "[...] open innovation likewise means that if one looks for such partners, not to constrain the search only to the own industry [...] that one actually has to think out of the box." (B3)

But even the most accurate search cannot always exclude the possibility of overlooking a suitable partner. Factors like globalization render it impossible to always have the full picture of the worldwide distribution of expertise:

> "The problem is, firstly, how do I actually know who is the world's most suitable? Perhaps I don't know that, because the world is complex enough. [...] There is a shortcoming according to the motto: 'You only know what you know'." (A4)

Even though a capable partner is elemental to the success of an OI-project, other aspects also come into play in the search and selection process. This is mainly because well-working co-operation not only depends on the partner's expertise but also to a significant extent on interpersonal factors. Therefore, relationship-related aspects are also factored into partners' selection:

> "The chemistry has to be right. Just as it works in every other team, [...] whether this is your working group at the university or whether this is a handball team or whether it is eventually such a project team, which in the end combines different [people] or employees of different companies." (B3; addendum by author)

Besides expertise and personal fit, the business model of the OI-partners is very relevant for partner selection. Since the clarification of IP rights can be a difficult task, companies tend to ensure OI-partners' business models are not too similar to the own one so as to avoid or reduce conflicts of interest regarding the allocation of such rights:

> "But what you, indeed, try to do is that you say: 'Ok, what are the interests of the single co-operators?' Meaning: company A, B, C – who has which interests. And as long as the business models of these co-operating companies are not too analog, there are actually few problems." (A3)

Companies can integrate academic partners, (e.g., universities, research institutes) and/or industrial partners (e.g., existing and new customers, suppliers, start-ups, consulting companies, competitors) in OI-projects. Generally, multiple partners can participate in one OI-project. However, co-operation tends to be bilateral. OI-projects involving customers in particular are often close to, or exclusively, one-to-one relationships, i.e., results of the OI-project may initially be used exclusively by the customer while acting as an OI-partner.

The results will possibly be offered to other customers only after the expiration of a deadline. Consequently, such OI-projects are part of wider strategic decision-making, since it presents a great opportunity to closely bind a respective customer to a company but also reduces the addressable market for this company:

> "Of course, every large customer wants to [...] have exclusive solutions. On the other hand, partially we cannot afford to give it completely exclusive to them, because the return of investment [...] is much too low then, which then does not cover the development costs and registration costs" (B1)

> "That means one definitely has to conceptually think within the company 'With whom do I want to do projects concerning what?' in order to not get tangles in some sort of maze, which later leads on the one hand to a legal uncertainty and, on the other hand, restricts you. So I think that is even one of the most critical issues that one needs to apply open innovation as a strategic element for the own company and, thus, must not limit one's view to the single OI-project and the single OI-activity in order to not get into massive difficulties." (A2)

The partner choice not only influences the number of participating parties. Several aspects of an OI-project might differ depending on the involved OI-partner (e.g., the project goals). However, not every difference between OI-projects is attributable to the type of OI-partner. Various aspects (e.g., past experience with an OI-partner, interchangeability of OI-partner) contribute to the heterogeneity of OI-projects, in the sense that also OI-projects conducted with the same type of partner can have distinguishing characteristics and so might differ from one other:

> "You also have to distinguish again between the types of external partners. [It is] something totally different to work in the sense of open innovation with an academic environment. [It is] something totally different to work in the sense of open innovation with a competitor, which one, indeed, sometimes does, or just work with a customer. Well, and then the quality of the co-operation also massively depends on the [...] relationship between company and customer – whether this is a key-account, i.e., a significant, frequent customer, or whether it is a prospective customer, whom I basically would like to turn into a customer through this OI-activity. These are totally different models." (A2; addendum by author)

Since an OI-project is characterized by multiple dimensions, it is difficult to make universal statements when only one dimension is studied. For instance, the expected quality of an OI-project cannot simply be derived from the type of OI-partner alone. There are positive as well as negative examples with all categories of partner:

> "We have good and bad examples from academia; and we have good and bad examples with start-up companies; and we have good and bad examples with other large industrial companies. [...] However, I did not see any statistics. But personally, I do not believe that there is statistical relevance in whether the one project is successful or the other is not successful." (A4)

Interviewees also provided different information based on the point of time when different OI-partners become involved in the innovation process. For example, the timing of customer integration is related to the partner's size and importance to the company. The larger and more important a customer, the sooner he/she will be involved in the innovation process so as to give him/her optimal influence:

> "Well, there are different stages. So I would say there is the idea phase; we only speak with the largest customers about the ideas. With projects that are in feasibility [phase], so to say the first proof of principle – also only with the largest customers. Than we have projects that are in the development phase. Those are typically one to three years away from the launch. [...] Then we speak with the regional and global customers. And everything, which is in the launch, is then basically available to all customers. That means there is a certain disclosure hierarchy of innovation." (B1, addendum by author)

With respect to academic and industrial partners, the difference is often between invention and innovation. Academic partners are primarily involved in the beginning of the innovation process where invention is at the forefront. Industrial partners, on the other hand, are often confined to the development phase, where the central task is to develop the invention into a saleable innovative product. At least in part, this different handling of academic and industrial partners is attributable to the fact that academic partners are more interested in fundamental research and less in the final functionality of an idea:

> "I mean that's the difference between invention and innovation. It is hard to innovate in university, but one can see if one pushes the enjoyment of inventing there and comes up with totally new ideas. As soon as it concerns innovation, i.e., the transformation of the idea or the invention into something saleable or marketable, the university is no longer the right place. And that is the difference, i.e., university is quasi discussing and seeing, whether one gets totally new ideas [...] into the process, whereas discussions with the [industrial] partners mostly run in the direction of innovation [...] The question of maturity – until one says we have solved a problem or a problem-solving approach – is [also] quite different. [...] A simple aspect is the scalability or the question of producibility. That concerns a scientist very little, while it is for us the key question at some point: 'Can we ever implement that?'" (A6; addendum by author)

In summary, the success of an OI-project is highly dependent on suitable partner choice. Firstly, it is important the expertise and the interests (especially regarding IP rights) of all participating partners are compatible. Additionally, a decent interpersonal relationship should exist or be built. In order to find an appropriate partner, companies often conduct a broad and extensive search, which is influenced by a range of project-specific aspects (e.g., project goals).

> "That is the reason why we do it [open innovation; annotation by author]. Absolutely worth it if the right partners work together." (A2)

5.4 Basic Conditions for OI-Projects

The right partner choice lays the foundation for a well-working OI-project. However, there are far more factors that play a crucial role in the success of open innovation. These factors are either related to the interaction and relationship of OI-partners or represent conditions that have to be fulfilled by each participant to make open innovation work.

As already indicated in chapter 5.2, an NDA and a clarified IP rights' allocation are the most important conditions. According to the interviewees, a solid contractual framework and the associated legal security are basic requirements for knowledge exchange in OI-projects. The contractual framework helps a company to protect its knowledge and to claim a proportion of any value generated through the OI-project. At the same time, it creates security for the employees involved in the OI-project and serves them as a guideline:

> "[...] the whole thing must be defined within a certain, reasonable contractual framework. Clear milestones have to be defined and also exit points." (B2)

> "The most important basic condition is a clear agreement concerning the IP rights, so that the employees know they can openly talk to the other colleagues, because the IP right situation is contractually well regulated. Based on experience, there are considerably less problems then." (A3)

An NDA is the minimum requirement and the most central component of this contractual framework. However, some OI-projects might call for the signing of additional contracts or agreements:

> "Well, what always belongs to the topic of open innovation – even though it might sound counterintuitive – is a non-disclosure agreement. What frequently goes with it is an MTA; material transfer agreement. What not always, but more than occasionally, goes with it is a joint development agreement [...] or right up to a strategic alliance agreement. These are simple contracts that regulate how knowledge and intellectual property are handled in this co-operation." (A2)

Besides regulating the allocation of IP rights, these contracts also record the milestones, reliability, and objectives of the OI-project. Not every OI-project aims to result in a patent application. However, if a patent is the defined project goal it is stipulated accordingly in the contractual framework. The final contracts are usually based on standard contracts, which have been adapted to the individual case, i.e., to the specific OI-project. The party preparing the first draft of the contracts therefore seeks to improve its bargaining position with respect to the final contract design:

> "And then it is generally advantageous if then the own lawyer party makes the first move. [...] This will be used as a basis for discussion and [...] you can imagine that if you made the proposal, then you are often in a somewhat better position as if you would be the one receiving the proposal." (B1)

Basic Conditions for OI-Projects

A second basic requirement for open innovation is trust between participating OI-partners. Mutual trust normally derives from factors such as positive experience with the partner and good reputation. However, a foundation of trust can also be built through contractual stipulations:

> "Trust certainly can be created by concluding appropriate, let's say, confidentiality agreements and similar things in contracts. But on the other hand, one certainly has to or certainly should have the feeling that certain findings, which were worked out during this project, that those are not immediately broadcasted to the whole world." (B3)

Another important condition is that all partners share similar attitudes and pull together during the OI-project. Open innovation can achieve results that are satisfactory for all participants only if all partners are willing to enter a balanced give-and-take relationship. Consequently, a common ground and fairness are also relevant elements for a successful OI-project:

> "The most important success factor is that all partners, who teamed up in this project, have the will to successfully finish the project. If there are some partners involved, who say: 'Well, let's see.' – That is always bad. [...] And if this is the attitude of the company [...], then you better stop the co-operation, because that is pointless. The company rather has to have a vital interest. [...] And this is then a sort out in the beginning; that you need to find out, what are we allowed to give so that the give-and-take is in balance. [...] And then the employees have to learn to understand this balanced give-and-take principle. That they take care, that they are at least not considerably more open than their counterpart." (A3)

There is a lack of formal structures to guide employees through an OI-project so they mainly depend on learning by doing. Consequently, employees need some experience with open innovation and the appropriate ways of working with external partners to be able to find the right balance between give and take:

> "My experience is they know very fast what they are allowed to say; what they are not allowed to say. There are some rules. Well, there is this rule 'one voice policy' [...] that means we do not issue any business data without consultation. But how to deal with people, [...] how confidential to be – that is something the people learn very quickly. [...] This is actually something they, again, get taught on the job by their team leader. So of course, there are various formal introductions, but they learn the real life through participating, imitating, or also through the instruction of the team leader." (A6)

They also have to cope with the delicate balancing act between openness and secrecy or IP protection because for the OI-project to be successful a company must not be too secretive or lose sight of its interest in IP rights:[117]

---

[117] This perspective is also supported by the literature. Hippel and Krogh 2003 demonstrate by using the example of open source software that companies can strike a balance between total openness and secrecy. Henkel 2006 refers to this balance as "selective revealing".

> "[A] scientist at the beginning of his career already has to be aware of the sensitivity of his statements, i.e., he already has to have a feeling for what [he] is allowed to say and what he is not allowed to say. If one completely buttons the lips and I am not allowed to say anything, it is not a good starting point. And if one starts to discuss his complete research project, it is also not quite healthy. And to find this balance between these two extremes, for me this is a bit a cultural question." (A6; addendum by author)

However, it is not enough for employees to know what to give. They additionally need to know the kind of information they can take without putting a possible patent application at risk:

> "Something that is always written down is the field, i.e., a description of the field in which a co-operation takes place. [...] And within this field you have to move. That is important, because the intellectual property subject is, of course, strongly tied up with it: 'What do I want to contribute? What am I not allowed to contribute? With what do I want to be – in inverted commas – contaminated? Now, with which knowledge also from the partner do I want to be contaminated?' "Contaminated" because afterwards, so to say, I cannot generate free background IP if this partner IP is verifiably integrated." (A2)

Another basic condition is that participants in an OI-project all speak the same language. This not only refers to a common mother tongue, foreign language or professional jargon, but rather emphasizes the importance of very clear and honest communication:

> "It is really important that both sides are quite clear then, because the worst would be if a customer makes me feel: 'Oh, yes. There is an enormous market. There is a huge need.' I leap at it with all my resources and after three years I realize: 'Well, it is not such great, terrific demand at all." (A1)

The contact between OI-partners was another aspect highlighted repeatedly by the interviewees. According to the respondents, contact should be frequent and face-to-face if possible, since many issues can be better clarified in person than by e-mail or telephone. Furthermore, personal contact enhances mutual trust.

Besides these requirements surrounding the interaction of OI-partners, certain internal conditions within each single company are also crucial for the success of open innovation. Management must have a favorable opinion toward open innovation and must support corresponding activities, i.e., there should be promoters (see Gemünden et al. 2007) within management who make sure employees can engage in OI-projects. Among other things, this support finds expression in making the necessary resources available and giving employees space to become involved in OI-projects and to discharge their related tasks:

> "If open innovation is truly lived and it reaches a certain critical mass, then appropriate resources must be provided for such a thing, of course." (B1)

> "The colleagues in clearly defined projects are under pressure to produce results; under time pressure; are not clear-headed, because they are simply embedded in operative objectives. They have to deliver. They hardly have any space. [...] Also the physical

closeness to operative units plays a role. If they are available – available on site – they are assigned and involved. Consequently, things that are set up in the sense of open innovation and new innovation fields, these colleagues have to be released for this and best case physically separated." (A5)

Another basic condition for successful OI-project is the equipping of employees with "absorptive capacity" (see Cohen and Levinthal 1990), i.e., employees need to have built a level of sufficient experience to be able to evaluate and assimilate external ideas. The relevance of absorptive capacity for open innovation is also highlighted repeatedly in the literature (see for example Lichtenthaler and Lichtenthaler 2009):

"What you need for the OI-process is, of course, you internally need a certain critical mass of intellectual know-how in order to be able to evaluate what is provided from outside. If you try to implement it within an evaluation-vacuum, than open innovation will more probably not be successful. Therefore, one can do blue sky open innovation if one has a lot of money. But if one is equipped with limited resources, then these investments must be rather goal-oriented." (B1)

In addition to absorptive capacity, other capabilities are crucial for open innovation. For example, employees have to be able to communicate with external partners and to drive the project forward. Furthermore, they have to have an open mind and certain sensitivity to other (corporate) cultures:

"That also means that one has to be a bit communicative. One has to have the ability to drive something forward together with external partners. One must be demanding on the one hand; loyal in dealing with external partners; and similar things, of course, come in addition." (B2)

"[...] capabilities in so far as the people have to be communicative, which not all of them are. They have to be open minded and willing to absorb things [...] ,i.e., they really must be open minded enough to overcome their inhibitions with respect to absorbing things that do not originate from them or that also may be in contrast to what they have done. They have to know the company well enough so that they can put the whole thing they learn in place. [...] Also the capability to overcome your own inhibitions concerning exchange with foreign cultures [is relevant]. (A6; addendum by author)

Employees have to be capable but they must also be willing to involve themselves in OI-activities. To encourage willingness, companies must create the appropriate framework conditions. The rationale for an OI-project should be clearly recognizable to employees involved in the specific project. This can be achieved through sufficient internal communication and it should also involve some personal benefit for the employee (e.g., fun, international experience). Finally, employees should also receive credit for their willingness to participate in OI-projects:

"The people must see meaning in it. [...] One has to have any personal benefit from it, in the sense that it has to be fun. One has to see a value, which one creates with it for the

company. And it also has to be encouraged in a sense that one at least gets a slap on the back and with the message: 'Well done!'" (A6)

Generally, open innovation requires a certain corporate culture, which promotes and supports exchange with external partners:

"It must be set up properly and has to take place in an appropriate environment, i.e., in the appropriate culture. [...] It requires the appropriate corporate culture – or let's say innovation culture – to be open for open innovation." (A5)

In summary, the choice of a suitable partner coupled with watertight legal agreements lay the foundation for OI-projects. Interviewees emphasized their belief that companies should give very high priority to their choice of partner(s) because this is a decisive factor for pooling all necessary expertise and enabling a trustworthy, open, and fair relationship. They also advocated a solid contractual framework, particularly with regard to clarifying the allocation of IP rights, as essential. A successful OI-project also demands that the relevant employees have certain abilities and a willingness to involve themselves in knowledge exchange. Above all, open innovation should be supported internally by management and a benign corporate culture.

5.5 Benefits and Challenges of Open Innovation

The most obvious reason for open innovation is that it can provide companies with expertise or technologies they do not have (cannot or will not develop) in-house. By importing another perspective it can help to stimulate the creativity of their employees and to help them think out of the box. Intelligent pooling of internal and external knowledge not only enables the acceleration of their own innovation processes, but also drives interdisciplinary issues – a phenomenon that appears to gain in importance. Interviewees noted a shift from standalone products to complex system solutions, which makes the pooling of competencies essential:

"I think times have changed here. We no longer make a product, which is then finished and sold, but it increasingly develops toward systems. It becomes more complex. It becomes more interdisciplinary." (A1)

Companies operating in sectors where innovations are usually capital-intensive and so, fraught with risk, can reduce or share this risk and spare their scarce resources by involving external partners:

"I am even firmly convinced that a company alone – if it doesn't want to put in billions in research – cannot raise this anymore. [...] So we have a project, where we work together with our competitors, because we simple say: 'That is so risky and still so far away.' – meaning seven to ten years. Here, we allow ourselves – at the level where we are today – to co-operate also with competitors. (A3)

"In the chemical industry, many innovations are rather capital- and, thus, time-intensive. [...] And therefore, open innovation itself is incredibly valuable for suppliers [...]" (B1)

This is particularly the case when companies operating in certain industries would like to develop into new, unknown, fields. If such a company strives for radical innovation or wants to develop new markets, than open innovation can considerably reduce the associated risk:

> "If you really want to do leap-frog research, really try something completely new, then meanwhile the thing is that each of these projects actually has to be done with a customer. Since otherwise the risk is really too high that you really [...] bind resources for three, four, five years and [...] develop past the market." (A1)

Another reason for open innovation is the striving for more efficiency. The involvement of external partners can help to enhance a company's internal innovation process and reduce the time to product launch:

> "[We] clearly recognized the trend of open innovation. The reason is that we are not the number one in the market. And if you are not number one, you usually do not have the same internal resources as the industry leader. That means, that you think about efficiency. You clearly can increase efficiency through expertise. You can increase expertise by sitting down with external partners [...] from university as well as industry or also from – we didn't do this yet – from NGOs." (B1; addendum by author)

To some extent, open innovation can also cultivate business relations (e.g., with customers) and obtain so-called "sticky information" (see Hippel 1994). For those companies active in B2B markets in particular, open innovation offers the additional positive flow-throughs from increased visibility and recognition by their peers, which may also enhance their future recruitment ability:

> "[...] you become more visible through these activities, because especially for companies like [us], which do not deliver to end customers, it is an advantage that should not be underestimated. [...] And HR is another topic, i.e., you get to know a lot of people and, that way, have a good source for the recruiting." (A6; addendum by author)

However, a company cannot realize all these benefits by ignoring the challenges associated with open innovation. Generally, the greatest challenge is to create the prerequisites for a successful OI-project (i.e., the IP situation and the partner choice, in particular). As already noted, companies have to master the balancing act between openness and IP protection. However, even an intensive and thorough partner search cannot rule out an element of risk – there is always the prospect the wrong one will be chosen. That said, the major and by far the most frequently mentioned challenge of open innovation is the additional administration and co-ordination required in OI-projects. For instance, the legal advisors responsible for drafting and evaluating all of the necessary contracts were cited as a significant bottleneck:

> "I think to some extent the largest bottleneck is – if you do a lot of open innovation – you need time to evaluate these contracts. It requires a certain experience for that. I possibly have a certain experience. The heads of laboratory will only partly have the experience, i.e., rely on legal resources then. [...] Of course, the priorities of CDAs versus a business

> *contract are very, very different, i.e., as a result you can arouse the head of laboratory's frustration if his priority – namely to get a CDA – is assessed relatively low and he has to wait for a long time." (B1)*

Consequently, extended decision-making processes and delays can occur. In addition, to certain extent companies in an OI-project partly cede control and become dependent on their partner's knowledge. Since frequent and open exchange is fundamental to OI-projects, knowledge outflow cannot be avoided. The challenge is to find the right balance between give and take:

> *"Disadvantage or a risk, which I certainly always take a little bit, is in the end that of course knowledge also flows out of the company, because I also have to […] contribute a certain degree of own knowledge in order to also smarten up my partner. Basically, I have to explain the problem more precisely." (B3)*

Another challenge is to neutralize unjustified negative attitudes toward external knowledge. Otherwise, the NIH-syndrome (see Clagett 1967 and Katz and Allen 1982) could cause employees to reject the external knowledge – with a consequent failure to integrate it into the internal innovation process:

> *"There is a certain arrogance-experience on both sides, i.e., the scientists believe they better understand science and the industrials think that they better understand the real life or economy, which is neither always true in this black-and-white painting. Then the not-invented-here problem […], i.e., if you externally find a topic that affects a working field, which is elaborated internally, there is naturally a defensive attitude from the subject owner." (A6)*

Considering open innovation as a strategic element also presents significant challenges. Open innovation will not always be the way for a company to go. For example, in case a customer strives for an exclusive OI-project, a company has to decide if it is economically worthwhile to agree on this co-operation or not. A final challenge mentioned by the interviewees is the different standards employed by various partners in OI-projects. These have to be reconciled. One such example cited was the different timescales expected in academia and industry:

> *"Then, of course, the time frames in the industry and in the university are quite different. So the university colleagues think that they at least have the time frame of a post-doc, in other words three years, or for a dissertation […], i.e., two, three years plus; while a typical project duration in the industry is one to two years." (A6)*

There are two sides to the OI-coin. While there are good reasons for involving external partners, there are also some potential disadvantages and obstacles to be addressed before deciding the feasibility of following an OI-approach. Nevertheless, the interviews show that overall the benefits outweigh the disadvantages. Interviewees were generally positive towards the concept. Based on the belief ambitious targets are no longer attainable under their own steam, their companies are wedded to the expansion of open innovation:

> "But the fact that we have so many co-operation projects certainly shows that we in net terms say: 'We accept the administrative effort and do it and we are also happy to do it.' Because in the end, the benefit is significantly higher than the effort. [...] If the expertise is not in-house, you only have the option not to do it or to do it externally." (A4)

According to the interviewed R&D managers, employees also are predominantly positive towards open innovation – at least as long as OI-projects are successful. R&D employees' eagerness to experiment entails a high intrinsic motivation to solve tricky and challenging tasks. Since open innovation is often applied in cases where the task is too complex and difficult for a single player, such projects offer the prospect of stimulating work and a certain degree of freedom, both of which are tremendous stimuli for R&D employees. They tend to focus on solving problems – regardless of whether they do it in-house or with external collaboration:

> "With respect to my employees, especially to my laboratory workers and so on, it is of course that they all are considerably self-motivated anyway. [...] a natural scientist himself likes to play around." (B3)

The better the personal fit with the external partner, the better the chances are that employees will buy into the process. In addition, past OI-experience has a substantial impact on the formation of employees' opinion, i.e., employees who have had previous positive experience of OI-projects will tend to be more positive about them than employees with little or no OI-experience. Further, the type of individual an employee is may also influence his/her preferences:

> "Let's say those, who are actively involved in OI-projects – I would say – do not have any preference. It always strongly depends on the success rate and on the experience from recent years – what's will be he outcome and does the whole thing make sense. Then, they are very open toward open innovation. [...] If somebody is rather introverted – without making a judgment here – introverted in the sense of he is an excellent researcher, but likes to work solitarily. Of course, he will have his problems with OI-projects, because it naturally also has a lot to do with communication and networking and demanding things and the like, whereas somebody, who has the inclination to do exactly those things, won't struggle." (B2)

Overall, the R&D managers believed their employees would only opt for closed innovation if they have had a negative experience with open innovation or if the results of a closed innovation process were as promising as an OI-process.

5.6 Chapter Summary

The statements of the R&D managers are predominately in line with the literature. Nevertheless, the interviews provide valuable insights into the practice of open innovation, highlighting in particular the neglected study of employees' perspectives towards open innovation. The interviews underline how important and sometimes difficult it is to find an appropriate partner for an OI-project and the crucial role the legal framework plays. In general, the managers' statements would imply the necessity for a wide range of OI-preconditions and strategic forethought. Accordingly, companies have to consider various aspects before opting for open innovation. As shown in Figure 15, there are good reasons to involve external partners but also disadvantages. Every company must decide whether the advantages outweigh the disadvantages on any individual project. Overall, the interviews show the participating companies are generally positive towards open innovation: The benefits clearly outweigh the negatives.

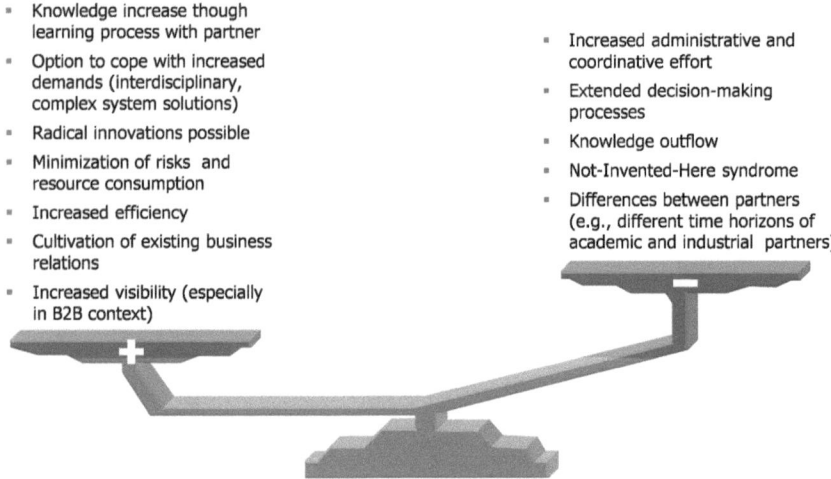

Figure 15: Advantages and Disadvantages of Open Innovation[118]

---

[118] Author's illustration

6 Findings from Quantitative Study (Online Survey)

This chapter summarizes the findings from the online survey.[119] The first sub-chapter gives some indications about data distribution and how biases were treated. It follows the description of the sample and some selected descriptive results. In the third sub-chapter, findings from an open-ended question regarding requirements for knowledge exchange in OI-projects are presented. Finally, I evaluate the measurement model and the structural model.

6.1 Data Distribution and Bias Treatment

In order to apply appropriate statistical tests and data analysis, the evaluation of data distribution was essential. Since data collection methods can create bias, it was also important to control for this factor.

6.1.1 Data Distribution

A first assessment was based on a graphical inspection of the data distribution and suggested non-normality for some of the items (cf. Hair et al. 2008, pp. 72f.). For the purpose of validating this first impression, skewness and kurtosis were examined (cf. Tabachnick and Fidell 2007, pp. 79f.). Following Osborne (2008a, p. 199), data can be considered normally distributed if the value of skewness and kurtosis do not significantly deviate from zero, i.e., the value of skewness and kurtosis should range between minus one and plus one. Since the values of many items were not in this range, the assumption of non-normality was further supported. For a final verification, normal distribution was tested by applying the Kolmogorov-Smirnov-Test and the Shapiro-Wilk Test (cf. Hair et al. 2008, p. 73; Weiber and Mühlhaus 2010, p. 147). Both tests confirmed the significant deviation of numerous latent variables from the normality assumption. Moreover, the Kolmogorov-Smirnov-Test and the Shapiro-Wilk were conducted on the construct level, demonstrating that several constructs show non-normality. Consequently, non-parametric tests had to be conducted. The suitability of the variance-based approach to estimate the structural equation model (i.e., the application of PLS for the data analysis) was also supported (see chapter 4.4.5.).

---

[119] The results of the online survey were partially published in Nedon and Herstatt 2014.

6.1.2 Bias Treatment

If respondents to a survey significantly differ from non-respondents, a non-response bias arises (cf. Sax et al. 2003, pp. 411f.). Following Armstrong and Overton (1977, pp. 397f.), I split each of the four company samples into first and last respondents and added all of the first respondents to one and all of the last respondents to another group. To check for non-response bias, I tested the answers of these two groups for any significant differences by applying a Mann-Whitney-U-test (cf. Bühl 2010, pp. 348ff.). Only one item (A3) showed a difference at a 5% level of significance. For all the other answers, no difference could be identified.

Common method bias is especially pervasive in behavioral research due to factors such as self-reporting and item characteristics but can be controlled in two ways, according to Podsakoff et al. (2003). The design of a study should aim to mitigate or avoid common method bias. Therefore, procedural remedies were applied before data collection. I followed the recommendations of Tourangeau et al. (2000) by using clear and consistent language, defining key terms (e.g., open innovation, knowledge exchange) at the beginning of the survey, and applying established items and measurement scales. Furthermore, I assured the employees of anonymity (cf. Podsakoff et al. 2003, p. 888). A second way of controlling for common method bias is by using statistical remedies – after data were collected – in order to minimize the effects of bias. In this study, I applied two statistical remedies. Firstly, I conducted Harman's single factor test, i.e., an exploratory factor analysis without rotation was applied to all items. When only one factor was extracted, this single factor explained only 22.27% of the variance, i.e., considerably less than half of the total variance. Furthermore, ten factors with eigenvalues greater than one were identified. Both results indicated the extent of variance, which cannot be attributed to the construct but to the measurement method, was not substantial (cf. Aulakh and Gencturk 2000, p. 529; Podsakoff and Organ 1986, p. 536). After conducting Harman's single factor test, I also checked the correlation matrix (see Table 15). The highest correlation was 0.511 and occurred between the intention to exchange documented knowledge (intention_doc) and the intention to exchange undocumented knowledge (intention_undoc). In cases of common method bias, very high correlations of above 0.9 would be expected (cf. Pavlou et al. 2007, p. 122). Therefore, the inspection of the correlation matrix did not provide any sign of common method variance. In summary, the questionnaire design as well as the tests conducted after the data collection suggests common method bias does not undermine this study.

6.2 Descriptive Results

As explained in chapter 4.4.4, a total of 283 R&D employees from four companies were asked to participate in the online survey. 199 R&D employees reacted to the request, providing 133 usable responses (see Figure 14). These 133 responses form the final sample. Thus, an overall response rate of 47% was achieved (see Table 7 for company-specific response rates). In the following, I will give an overview of the sample with respect to demographics, employees' company-related details, and OI-experience. Thereafter, I aim to highlight the most interesting descriptive results from the survey.

On average, respondents were 42.3 years old and predominantly male.[120] Only 18.0% of the R&D employees were female. The overwhelming majority said they held a university degree and had graduated in the fields of natural science or engineering.[121] Only 10.0% did not have a higher education and fewer than 5.0% held degrees in fields other than natural science or engineering. Respondents were mostly located in a German office and had been employed for 13 years at their respective companies.[122] These structural characteristics also applied to the sub-samples of the four participating companies, which showed only marginal differences from one another. Table 8 illustrates total sample and sub-sample characteristics.

Although the sample seemed to be unbalanced and homogeneous at a first glance, a closer look showed it was very representative of R&D departments. A tertiary education in the field of natural science or engineering is often a job requirement in R&D departments. This is very well reflected in the educational background of the surveyed R&D employees. However, as stated by the Statistische Bundesamt (cf. Mischke and Wingerter 2012, p. 22), women are strongly under-represented in these fields of study. Consequently, my sample can be considered representative in an R&D context.

Respondents worked on average on 4.7 OI-projects during the last three years and on 9.2 OI-projects during the last ten years (see Table 8). As Figure 16 shows, most experience arose from OI-projects carried out with universities and/or research institutes.[123] Considering the characteristics of the four surveyed companies (e.g., commitment to open innovation, high levels of internal R&D), this result is not surprising and in line with the literature (see Laursen and Salter 2004; Tether and Tajar 2008). Customers were the second most popular

---

[120] The average age was calculated based on 127 employees, who (correctly) stated their age. The share of male and female respondents was calculated based on 128 employees, who stated their gender (see chapter 4.4.4).
[121] The share of PhD, master/diploma, bachelor, or apprenticeship was calculated based on 130 employees, who stated their highest educational degree. The share of natural science, engineering, and/or economics was calculated based on 128 employees, who stated their field of education (see chapter 4.4.4).
[122] The share of different locations was calculated based on 130 employees who stated their location. The average tenure was calculated based on 129 employees who (correctly) stated their tenure (see chapter 4.4.4).
[123] Respondents could evaluate each OI-partner independently, e.g., an employee could state if he/she had worked together very often/very rarely with all five partners mentioned.

OI-partner, followed by industrial partners (excluding suppliers and competitors). Collaboration on OI-projects with competitors was rare but it did sometimes occur.

Table 8: Sample and Sub-Sample Characteristics

		Total Sample	Company A	Company B	Company C	Company D
Responses	*(usable)*	133	58	33	35	7
Age	*(average)*	42.3 years	42.0 years	43.4 years	41.9 years	42.0 years
Gender	Male:	82.0 %	83.3 %	65.5 %	93.9 %	83.3 %
	Female:	18.0 %	16.7 %	34.5 %	6.1 %	16.7 %
Highest Degree	Apprenticeship:	10.0 %	0 %	36.4 %	2.9 %	0 %
	Bachelor:	1.6 %	3.6 %	0 %	0 %	0 %
	Master/diploma:	29.2 %	14.3 %	9.1 %	67.7 %	57.1 %
	PhD:	59.2 %	82.1 %	54.5 %	29.4 %	42.9 %
Field of Education	Natural science:	61.7 %	87.3 %	90.6 %	2.9 %	14.3 %
	Engineering:	33.6 %	7.3 %	3.1 %	94.2 %	85.7 %
	Economics:	4.7 %	5.4 %	6.3 %	2.9 %	0 %
Tenure	*(average)*	13.0 years	11.0 years	15.7 years	14.0 years	11.3 years
Location	Germany:	82.3 %	66.1 %	87.9 %	100.0 %	100.0 %
	Europe (rest):	6.2 %	12.5 %	3.0 %	0 %	0 %
	Brazil:	9.2 %	19.6 %	3.0 %	0 %	0 %
	Others:	2.3 %	1.8 %	6.1 %	0 %	0 %
Number of OI-Projects	Last 3 years	4.7	5.8	4.8	2.7	5.1
	Last 10 years	9.2	10.0	11.1	6.2	9.0

Figure 16: OI-Partners (OI-Experience)[124]

For the company presentations and follow-up group discussions (see chapter 4.1), all survey questions were analyzed and presented on a descriptive level. At this point, only the most interesting findings are presented.

---

[124] Author's illustration

The first insight relates to employees' intention to exchange knowledge in OI-projects. Respondents indicated they were more likely to exchange undocumented knowledge with external partners than to share documented knowledge within OI-projects (see Figure 17). This finding was also statistically tested and verified and is further supported by the literature (see Constant et al. 1994).[125] Furthermore, three possible reasons for this difference were identified during the three follow-up discussions:

1) Confidence – *"Everything that is carved in stone, I could be held responsible for".* [126] Documented knowledge can be stored and cited as evidence that something has been communicated. In contrast, verbal exchanges don't leave a paper trail. Employees do not want to exchange information that might go beyond the negotiated NDAs and most of *"[...] the people are well trained on what knowledge can be shared and what cannot. Therefore, they are more careful with the documented stuff."*[127]

2) Efficiency – *"It is easier and faster to pick up the phone than to write an e-mail."*[128] Employees can avoid the effort to collect all relevant documents or write everything down. Moreover, some information might be hard to document or would *"[...] convert a ten-page document into a 100-page document."*[129] Therefore, employees might prefer undocumented knowledge.

3) Information quality – *"Undocumented knowledge is more up to date, since the documentation needs time".*[130] Undocumented knowledge is, therefore, *"[...] often more valuable, because it is the latest information. With documented knowledge, one has to assume that it has already been shared with others. A document is quickly distributed. Additionally, some things might only develop from conversations."*[131] Last but not least, past experience might have shown that undocumented knowledge is more valuable and, therefore, preferable.

---

[125] The influence of attitude, subjective norm and perceived behavioral control on the intention to exchange documented knowledge in OI-projects was modeled, i.e., intention was operationalized solely based on the documented knowledge-related items I1 and I2 (see Table 4). In this case, attitude, subjective norm, and perceived behavioral control explained 32% of the variance ($R^2 = 0.320$). Secondly, the influence of attitude, subjective norm, and perceived behavioral control on the intention to exchange undocumented knowledge in OI-projects was modeled, i.e., intention was operationalized solely on the basis of the undocumented knowledge-related items I3, I4, and I5 (see Table 4). In this case, attitude, subjective norm, and perceived behavioral control explained 46% of the variance ($R^2 = 0.459$). In addition, the path coefficients were stronger and on a higher significance level.

[126] This citation was taken from follow-up group discussion with R&D representatives of Company B (February 12, 2013). The original citation was in German and translated into English. This is also true of all the following citations from the follow-up group discussions.

[127] Citation from follow-up group discussion with R&D representatives of Company A (February 22, 2013).

[128] Citation from follow-up group discussion with R&D representatives of Company B (February 12, 2013).

[129] Citation from follow-up group discussion with R&D representatives of Company B (February 12, 2013).

[130] Citation from follow-up group discussion with R&D representatives of Company A (February 22, 2013).

[131] Citation from follow-up group discussion with R&D representatives of Company B (February 12, 2013).

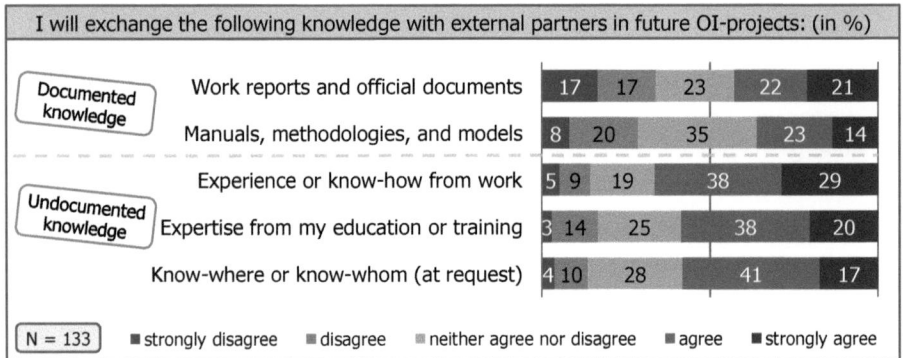

Figure 17: Descriptive Results regarding Intention[132]

A second finding concerns employees' attitude toward exchanging their knowledge with external partners in OI-projects. As can be seen in Figure 18, where the percentage of all positive replies is displayed, the vast majority of the surveyed R&D employees have a (very) positive attitude toward their knowledge exchange with external partners. 99% find the knowledge exchange with external partners (very) valuable for themselves and 86% think it a (very) pleasant experience. This implies the NIH-syndrome (reflecting a negative attitude toward external input) may not play a relevant role among the respondents.

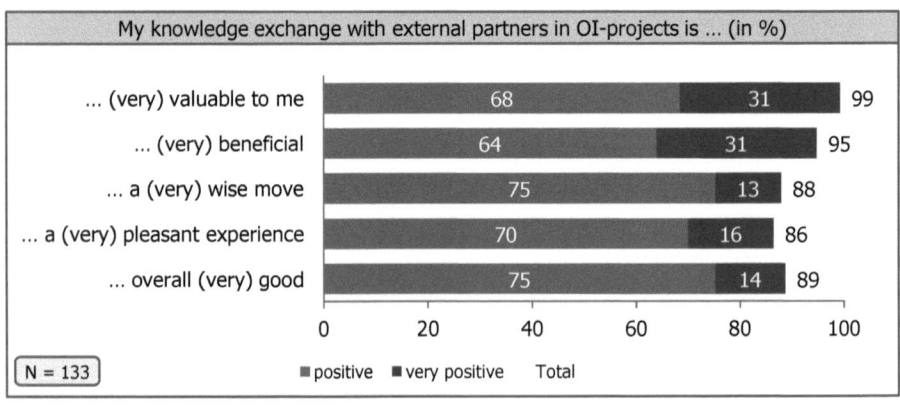

Figure 18: Descriptive Results regarding Attitude[133]

---

[132] Author's illustration
[133] Author's illustration

The third descriptive finding is related to subjective norm, the only formatively measured construct of this study. As shown in Figure 19, employees believe their immediate supervisor most wants them to exchange knowledge in OI-projects; followed by the CEO and their colleagues. However, the CEO carries the greatest weight.

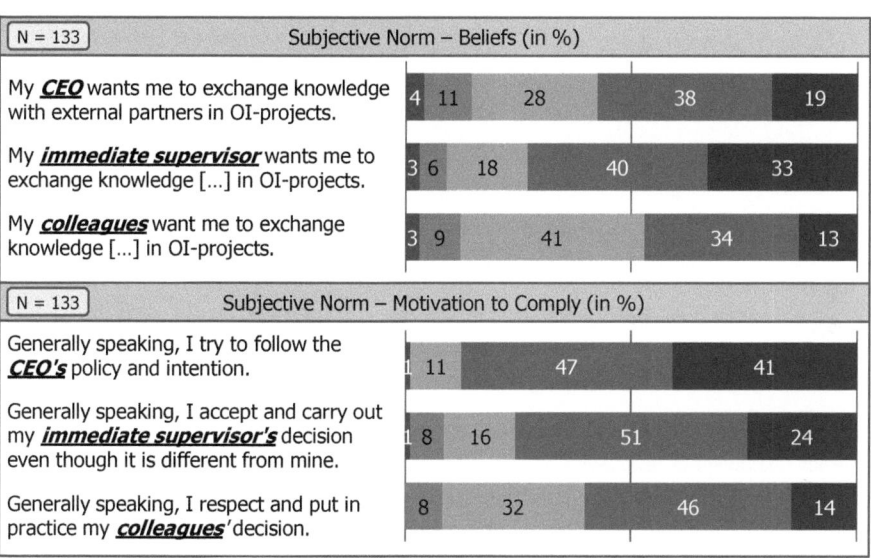

Figure 19: Descriptive Results regarding Subjective Norm[134]

A fourth insight, which is worth highlighting, is related to rewards. As described in chapter 4.4.1.8, five items were added after the pre-test. The descriptive data analysis displayed in Figure 20 confirms the statements of the pre-testers and shows employees are much more motivated by rather intrinsic rewards (pre-test items) than by organizational, rather extrinsic rewards (original items).

---

[134] Author's illustration

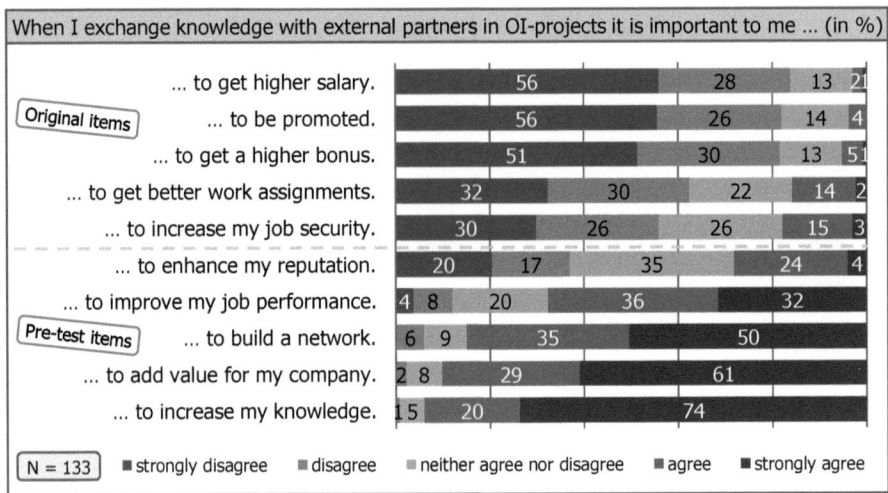

Figure 20: Descriptive Results regarding Rewards[135]

A last descriptive finding is related to the sense of self-worth and presented in Figure 21. Generally, knowledge exchange in OI-projects creates a certain sense of self-worth. However, the degree depends on the considered outcome. The vast majority of the employees surveyed considered their involvement in OI-projects valuable in creating new business opportunities for their organization. However, only few believe it improved work processes within their company in any significant way.

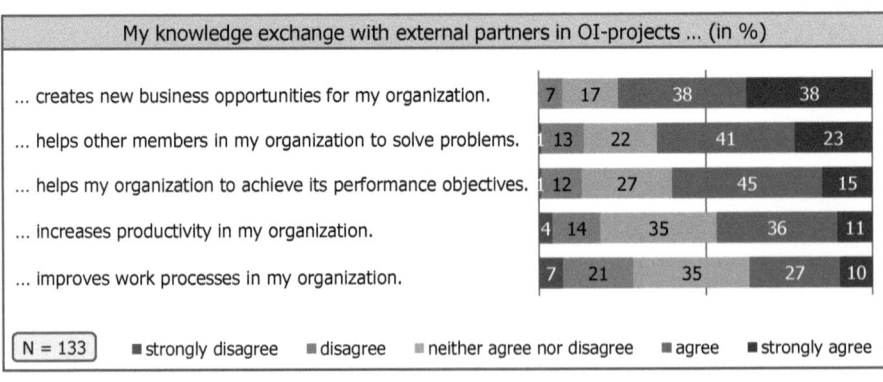

Figure 21: Descriptive Results regarding Sense of Self-Worth[136]

---

[135] Author's illustration
[136] Author's illustration

6.3 Findings from an Open-Ended Question

The insights resulting from the only open-ended question in the survey relate to basic conditions for knowledge exchange in OI-projects. Respondents were asked to state up to five requirements that must be met to enable them to exchange knowledge with external partners in OI-projects. Even though not all respondents answered this voluntary question, 390 requirements were listed. Most of the answers consisted of less than five words. To analyze the replies, a general inductive approach as diagrammed in Figure 22 was conducted (see Thomas 2003).

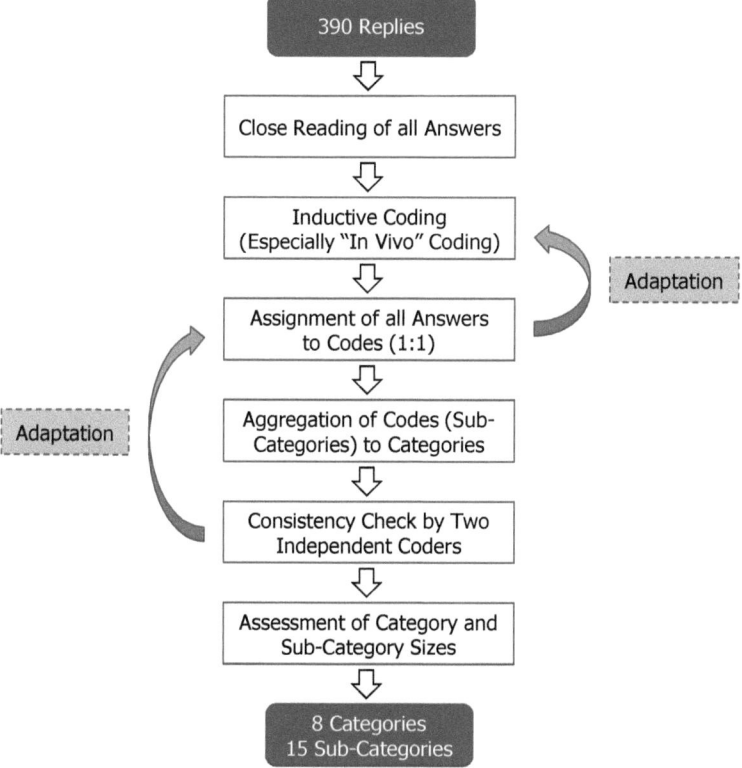

Figure 22: Research Approach for Open-Ended Survey Question[137]

The first thing I noted from scrutinizing the replies was that some were repeated verbatim.[138] In these cases, the expression of the respondents was adopted as category

---

[137] Author's illustration
[138] The words NDA, IP, and trust were mentioned multiple times.

name, i.e., in vivo coding was applied (cf. Strauss 1987, pp. 33f.). If the category was not as obvious, codes were derived by reading the answers repeatedly. However, the wording of these categories was also adopted from appropriate expressions of respondents.

After the first draft of a coding system had been developed, I tried to assign each reply to one of the categories.[139] This process unveiled some distracting overlaps between the categories; adaptation of the coding system was necessary and answers had to be reassigned.[140] Once this iterative procedure was finished, 355 (out of 390) replies were assigned to one of 15 categories. The remaining 35 answers were either labeled "miscellaneous" (28 cases) or "not understandable" (seven cases). When all replies were labeled with an appropriate code, I aggregated the 15 categories to eight higher-order categories to get a better overview (see Figure 23).

As a next step, a consistency check was conducted. I asked two independent evaluators to assign the 390 replies to one of the 15 sub-categories or to label them "miscellaneous" or "not understandable".[141] By doing so, the evaluators also assigned the replies to one of the eight aggregated categories. After the evaluators had finished the classification and submitted their results, I compared my coding with theirs. The first evaluator assigned 301 replies (77.2%) to the sub-category that I had chosen. The second evaluator did so with 305 replies (78.2%). In 256 cases, both evaluators agreed with my coding and one of them also approved my decision in 77 additional cases. Consequently, a consensus about the coding with at least one of the evaluators was reached in 333 and respectively 85.4% of the cases, so that the data and its codification could be considered reliable (cf. Taylor and Watkinson 2007, p. 53). Thus, the coding scheme and the original assignment of these 333 replies were retained. However, in 57 replies the coding had to be reconsidered. In 25 of the 57 cases, both evaluators had selected the same sub-category. Here, the original coding was changed and the evaluators' classification was adopted. During a follow-up discussion with one of the evaluators, we agreed to adapt the coding of further 18 replies.[142] In 14 cases, no consensus could be reached, which equated to 3.6% of the 390 replies. Due to this minor discrepancy between the evaluators and the researcher, the difference was considered negligible, i.e., no adjustments were made in this instance. Since the consistency check was considered successful, a descriptive analysis was then conducted.

---

[139] Since a descriptive analysis of the replies was intended, each reply was limited to a single code, i.e., it had to be a one-to-one relationship.
[140] Despite every effort, it was not possible to eliminate overlaps completely. Therefore, categories were aggregated to more distinctive higher-order categories.
[141] The two evaluators were instructed to assign each reply to one of the 15 sub-categories, or to label them as "miscellaneous" or "not understandable". They received the list with all 390 responses, the names of the 15 sub-categories and the requisite information on how the 15 sub-categories relate to the 8 categories.
[142] I changed six of my codes and the evaluator reassigned 12 replies. Even though the 18 replies were assigned to new sub-categories, the higher-order category mostly remained the same.

For that purpose, the number of assigned replies per (sub-) category was assessed. The results for the main category are displayed in Figure 24.

Sub-Categories	Categories
1) Confidentiality, NDA	
2) Clarification of IP and Exploitation Rights	1. Legal Security
3) Clear Legal/Contractual Framework	
4) Trust	
5) Good Rapport and Communication (Open, Frequent, Personal)	2. Good Rapport
6) Fairness (Give-and-Take), Win-Win Situation	
7) Common Goals and Interests	3. Common Ground and Fairness
8) Expertise	
9) Complementary Capabilities	4. Expertise
10) Clarity about Objectives and Project Scope (Tasks, Milestones, etc.)	5. Clarity
11) No Rivalry	
12) Appropriate OI-Partner	6. General Partner-Fit
13) Foreseeable Success/Benefit of OI-Project	7. Added Value
14) Freedom of Action (Time, Money, etc.)	
15) Management Support	8. Freedom of Action

Figure 23: Categories of Requirement for Knowledge Exchange in OI-Projects[143]

Nearly one third of all analyzable answers[144] related to legal security, i.e., almost all respondents pointed out the need for NDAs, agreements on IP, and/or a contractual framework. 21% of the replies referred to a good rapport and deep relationship characterized by mutual trust and an open, frequent, and – at best – personal dialogue. Common ground and fairness were also considered among the most important requirements for knowledge exchange in OI-projects. Respondents emphasized the necessity of making it a win-win situation for all parties participating in an OI-project – common goals and interests are essential. 10% of the 355 answers related to OI-partners' expertise. According to the surveyed employees, a partner should be an expert in the relevant area and bring in complementary know-how.

---

[143] Author's illustration
[144] Answers labeled "miscellaneous" or "not understandable" were excluded, i.e., the final sample is N = 355.

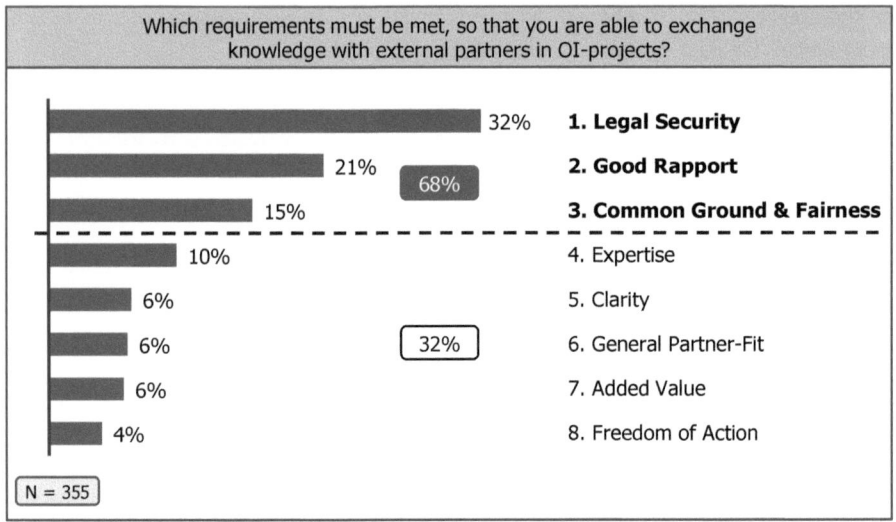

Figure 24: Requirements for Knowledge Exchange in OI-Projects[145]

Clarity, general partner-fit, and added value were mentioned relatively often and accounted for 6% each of the overall responses. Clarity refers to the requirement that project scope (including tasks, project objectives, timeline, budget, and next steps) have to be clearly defined. General partner-fit means the partner has to be reliable and rivalry must be avoided, i.e., direct competitors are seldom preferred partners. Furthermore, the OI-project should add value that other or "closed" innovation projects cannot accomplish. Lastly, 4% of all replies were concerned with the need for freedom of action, i.e., management support is required to ensure the necessary resources are available to make experimentation possible.

6.4 Measurement Model

In order to evaluate whether items measured their assigned construct properly, the measurement model (outer model) of all constructs had to be assessed (cf. Hair et al. 2012, p. 423). However, the evaluation of reflective and formative measurement models is based on different criteria (cf. Hair et al. 2011, pp. 145ff.; Henseler et al. 2009, pp. 298ff.; Henseler et al. 2012, pp. 269ff.). Thus, it is conducted separately in the following.

6.4.1 Reflective Constructs

To optimize the assessment of the reflective measurement models, a first and second generation method was applied (cf. Hair et al. 2014, p. 2). An exploratory factor analysis was

---

[145] Author's illustration

conducted as a first generation method, followed by a confirmatory factor analysis, representing a second generation method.

6.4.1.1 Exploratory Factor Analysis (EFA) – A First Generation Method

The intent of an EFA primary is to identify or explore the underlying structure among items and to summarize or reduce data (cf. Hair et al. 2008, pp. 94ff.; Netemeyer et al. 2003, p. 121). Thus, it is not necessarily applied in structural equation modeling because this multivariate method is usually based on conceptual considerations regarding the underlying structure, i.e., the number of items and items' affiliation to the constructs (cf. Grote 2010, p. 118).[146] However, EFA can be combined judiciously with other multivariate techniques such as structural equation modeling (cf. Hair et al. 2008, p. 100), as it is a first generation method to check for reliability and validity of reflective measures (cf. Herzog 2008, p. 138; Weiber and Mühlhaus 2010, pp. 104f.).[147]

In this study, an EFA was conducted for five reasons: Firstly, it was intended to show if perceived behavioral control comprises two independent factors (perceived self-efficacy and perceived controllability) as suggested in chapter 3.1.3. Secondly, it was used to demonstrate yet again the wisdom of differentiating between the intention to exchange documented knowledge and the intention to exchange undocumented knowledge (see chapter 6.2.). A third reason refers to the items measuring rewards. As described in chapter 4.4.1.8 and 4.4.2, five additional items that came up during the pre-test were added to the original five reward items from Kankanhalli et al. (2005). The EFA was conducted to check whether all items are highly interrelated and so constitute only one factor or if the ten items measure multiple factors (cf. Hair et al. 2008, p. 94). Fourthly, the EFA was conducted to review all reflectively measured, model-relevant constructs. Even though existing and repeatedly employed items and constructs were used in this study (implying that an EFA is not necessary), they had not yet been applied in the context of OI-research. It was, therefore, considered beneficial to execute the EFA in order to see if only the planned constructs would be extracted; all items would load on the intended constructs; and the exclusion of single items would improve the measurement model of a specific construct (cf. Weiber and Mühlhaus 2010, p. 106). However, the final decision about excluding an item was not solely based on the EFA results, but also on the outer loading[148] relevance testing recommended by Hair et al. (2014, p. 104). The last reason to conduct an EFA is related to

---

[146] According to the decision diagram of Hair et al. 2008, p. 97, a confirmatory factor analysis (CFA) is the current best method in structural equation modeling. In chapter 6.4.1, the CFA was conducted to confirm the measurement theory (cf. Hair et al. 2008, p. 693). By combining the bottom-up (EFA) and top-down (CFA) factor analysis approach, the fit of the measurement model was tested optimally.
[147] To evaluate the reliability, only one construct is included in the EFA. When assessing validity, all constructs have to be considered (cf. Weiber and Mühlhaus 2010, p. 104).
[148] Outer loading and indicator loading are equivalent terms and used interchangeably. However, indicator loading is mostly used in the following.

the control variable used to measure the "Big Five" personality traits of a person with ten items. The EFA had to clarify which of the five personal traits could correctly be extracted (cf. Weiber and Mühlhaus 2010, p. 106).

To address these five issues, the EFA was conducted collectively for all model-relevant constructs (overall EFA) and for each construct separately. Principal component analysis with promax rotation was employed for the extraction of factors (cf. Hair et al. 2008, pp. 105ff.; Nunnally and Bernstein 1994, pp. 491ff.).[149] Kaiser-Meyer-Olkin-criterion (KMO-criterion) and Bartlett's test of sphericity were used for testing the assumptions for factor analysis. The KMO-criterion was required to exceed the threshold 0.5 (cf. Kaiser and Rice 1974, pp. 112f.) and the significance of Bartlett's test of sphericity had to be below 0.05 (cf. Dziuban and Shirkey 1974, pp. 358ff.; Hair et al. 2008, pp. 104f.) to fulfill the requirements for factor analysis. For the overall EFA, the KMO-criterion was 0.761 and the significance of Bartlett's test of sphericity was zero, which both suggested the general appropriateness of the correlation matrix for factor analysis application (cf. Backhaus et al. 2011, p. 341; Dziuban and Shirkey 1974, pp. 358ff.; Hair et al. 2008, p. 105). With respect to single, model-relevant constructs[150], the KMO-criterion ranged from 0.604 to 0.786, also indicating the suitability of using EFA (cf. Backhaus et al. 2011, pp. 342f.; Kaiser and Rice 1974, pp. 112f.). Furthermore, the significance of Bartlett's test of sphericity was always zero, again implying sufficient correlations among the items for conducting EFA (cf. Dziuban and Shirkey 1974, pp. 358ff.; Hair et al. 2008, p. 105). The results of the overall EFA are displayed in Table 9. The outcome of the EFA applied separately for each construct can be seen in Tables 10 - 12.

To assess and interpret the EFA results, widely accepted cut-off values for measure of sampling adequacy (MSA), factor loadings (= component loadings if principal component analysis is applied as an extraction method[151]), communality, and variance extracted were consulted. Following Backhaus et al. (2011, p. 341) and Hair et al. (2008, p. 104), the MSA value had to be at least 0.5. The requirements regarding factor loadings depend on sample size. Given the sample size of this study (N = 133), a factor loading of at least 0.5 was necessary to be deemed significant, according to Hair et al. (2008, p. 117). In actual fact, only items with a factor loading above 0.6 were considered satisfactory in this study. Communality is generally suggested to be at least 0.5 (cf. Hair et al. 2008, p. 136; Weiber

---

[149] Principal axis factoring is said to yield the best results if the distribution is significantly non-normal (cf. Osborne et al. 2008, p. 89). Nevertheless, the widely used principal component analysis has been applied because one of the main reasons for conducting the EFA was data reduction (cf. Hair et al. 2008, pp. 107f.). To control for differences resulting from the employed method, principal axis factoring was also conducted. The comparison of the results shows both approaches suggest the same conclusions. With respect to the rotation of factors, an oblique method (promax rotation) was chosen, since the basic assumption (that factors are correlated to some extent) is more realistic than assuming uncorrelated factors, which is the hypothesis of orthogonal rotations (cf. Hair et al. 2008, p. 116; Osborne et al. 2008, p. 90; Weiber and Mühlhaus 2010, p. 107f.).

[150] The KMO-criterion of the control variable assessing the "Big Five" was 0.590 and the significance of Bartlett's test of sphericity was zero.

[151] In the following, factor loading and component loading are equivalent terms and used interchangeably.

and Mühlhaus 2010, p. 170). However, as the *"[c]ommunality is the sum of squared loadings (SSL) for a variable across factors"* (Tabachnick and Fidell 2007, p. 621) and a factor loading of 0.5 was sufficient in this study, items with a communality of at least 0.25 (= 0.5^2) could be considered appropriate if the item loads on only one factor. Therefore, items with communality lower than 0.5 were not deleted. In fact, a 0.4 cut-off value was appointed. Lastly, the variance extracted had to exceed 0.5 (i.e., 50%), as suggested by (Fornell and Larcker 1981, p. 46).

In addition to MSA, factor or component loadings, communality, and variance extracted; inter-item-correlation (IIC) and corrected item-total-correlation (ITC) were taken into account. Following Du Preez et al. (2008, p. 62) and Robinson et al. (1991, p. 13), the benchmark level for IIC was set at a minimum of 0.3. The value of ITC had to be at least 0.3 (cf. Blankson and Kalafatis 2004, p. 18; Du Preez et al. 2008, p. 62), but, optimally, above 0.5 as suggested by Bearden et al. (1989, p. 475) and Zaichkowsky (1985, p. 343). Furthermore, key figures of the confirmatory factor analysis (indicator loadings, average variance extracted (AVE), composite reliability/Dillon-Goldstein's ρ, and Cronbach's alpha) were consulted to validate the EFA-based decision on item exclusion (cf. Hair et al. 2014, pp. 103f.). A detailed examination of these figures will follow in chapter 6.4.1.2.

a) EFA Applied to Attitude

During the overall EFA (see Table 9), the attitude construct was correctly identified, i.e., all items intended to measure attitude showed the highest loading on the same component. When the EFA was conducted separately for the construct (see Table 10), the communality of item A1 and the variance extracted did not meet the requirements, which suggested excluding item A1 (cf. Hair et al. 2008, pp. 119f.; Weiber and Mühlhaus 2010, p. 109). This was further supported by the scrutiny of its indicator loading (see Appendix C) and the impact of its exclusion on the construct's AVE and composite reliability (cf. Hair et al. 2014, p. 104). The indicator loading of item A1 was below the threshold value of 0.7 and the exclusion would improve the AVE considerably. Although another item (A4) had a lower indicator loading than item A1, the elimination of A1 was preferable because its exclusion showed a stronger positive effect on AVE and a lesser negative impact on composite reliability. By excluding item A1, the variance extracted of the attitude construct was raised to 51.7% (cf. Table 10 and Table 13) and the AVE was increased above the threshold of 0.5 (cf. Appendix C and Table 14).

Table 9: MSA, Communalities and Pattern Matrix – Overall EFA

	MSA	Comm-unality	\multicolumn{10}{c}{Component}
			1	2	3	4	5	6	7	8	9	10
Critical Value	> 0.5	> 0.4	\multicolumn{10}{c}{Loading > 0.5}
A1	0.832	0.583	**0.631**	-0.246	-0.071	0.390	-0.100	-0.033	0.161	-0.009	-0.121	-0.123
A2	0.768	0.523	**0.430**	-0.009	-0.259	0.039	0.311	0.027	-0.110	-0.046	0.182	0.121
A3	0.886	0.560	**0.615**	0.105	0.062	-0.096	0.082	0.085	-0.100	0.128	0.141	-0.034
A4	0.755	0.702	**0.707**	0.139	0.031	-0.311	0.110	0.077	0.130	-0.109	-0.027	0.397
A5	0.803	0.569	**0.722**	-0.115	0.120	0.139	0.174	-0.177	-0.027	0.097	-0.068	-0.039
PBC1	0.697	0.715	-0.071	**0.802**	-0.193	0.067	-0.172	-0.071	0.080	-0.041	0.077	0.251
PBC2	0.705	0.700	0.169	**0.770**	-0.105	0.161	-0.169	-0.135	-0.138	0.136	-0.027	0.053
PBC3	0.873	0.660	-0.136	**0.824**	0.090	0.134	-0.008	0.052	-0.031	0.124	-0.170	-0.032
PBC4	0.756	0.789	-0.052	**0.841**	0.179	-0.164	0.104	-0.026	-0.010	-0.132	0.073	-0.078
PBC5	0.738	0.783	-0.012	**0.793**	0.256	-0.189	0.007	0.013	0.015	-0.149	0.135	-0.013
PBC6	0.870	0.564	0.264	0.263	0.064	0.381	-0.041	0.008	0.013	-0.085	-0.158	**0.428**
I1	0.765	0.734	0.014	0.095	**0.790**	0.271	-0.036	-0.090	0.071	0.001	0.122	0.125
I2	0.734	0.598	0.126	0.010	**0.591**	0.427	0.151	0.034	-0.142	-0.033	-0.021	0.071
I3	0.834	0.646	-0.039	0.208	0.278	**0.674**	0.086	0.017	-0.072	0.112	-0.118	0.051
I4	0.824	0.712	0.070	-0.068	0.100	**0.830**	-0.083	-0.033	-0.016	-0.070	0.161	-0.028
I5	0.820	0.729	-0.037	-0.034	0.246	**0.826**	0.036	-0.031	-0.023	-0.120	0.116	0.048
JOY1	0.791	0.692	0.181	0.076	0.001	-0.057	**0.793**	0.059	-0.116	-0.077	-0.007	-0.127
JOY2	0.703	0.811	0.137	-0.116	0.056	0.002	**0.862**	0.062	-0.005	-0.079	0.037	0.092
JOY3	0.648	0.627	-0.029	-0.138	0.009	0.055	**0.743**	0.039	0.146	0.027	-0.110	0.035
SW1	0.827	0.526	-0.020	0.037	-0.221	0.219	0.106	**0.518**	-0.022	-0.061	0.117	-0.091
SW2	0.750	0.515	0.025	0.081	0.138	-0.016	0.080	**0.537**	0.147	0.100	-0.213	-0.248
SW3	0.681	0.728	-0.179	-0.217	-0.022	0.098	0.050	**0.772**	-0.206	0.004	0.208	0.287
SW4	0.711	0.754	-0.040	-0.034	-0.062	-0.128	0.111	**0.897**	-0.014	0.001	0.001	0.084
SW5	0.678	0.714	0.271	0.014	0.066	-0.045	-0.199	**0.568**	0.073	0.092	0.186	-0.240
RP1	0.890	0.557	-0.013	**0.358**	-0.149	0.179	0.120	0.241	0.197	0.203	-0.030	-0.034
RP2	0.678	0.746	0.087	-0.021	0.002	-0.188	-0.007	-0.044	**0.872**	0.037	0.047	0.062
RP3	0.656	0.810	0.010	-0.052	0.085	-0.026	0.005	-0.082	**0.917**	-0.071	0.067	0.030
RP4	0.822	0.683	-0.080	0.013	-0.128	0.261	0.036	0.110	**0.709**	-0.074	-0.029	0.131
REW1	0.687	0.584	-0.120	-0.026	0.125	-0.040	0.373	0.096	0.115	**0.522**	-0.126	0.153
REW2	0.808	0.845	0.099	-0.074	-0.047	-0.023	-0.059	0.069	-0.035	**0.928**	-0.060	0.031
REW3	0.717	0.868	-0.003	0.045	-0.007	-0.072	-0.054	0.040	-0.085	**0.945**	0.000	0.138
REW4	0.730	0.867	0.066	0.056	-0.048	-0.043	-0.157	0.037	-0.033	**0.950**	0.016	0.116
REW5	0.670	0.690	-0.071	0.007	0.131	0.001	0.021	-0.011	0.110	0.364	0.217	**0.700**
REW6	0.777	0.690	0.018	-0.096	0.058	0.032	0.175	-0.378	0.092	**0.527**	0.381	0.034
REW7	0.692	0.587	0.021	0.235	-0.258	0.064	0.231	-0.094	-0.055	0.073	**0.428**	-0.122
REW8	0.791	0.786	-0.164	0.163	-0.053	0.150	0.234	-0.012	0.102	-0.075	**0.633**	-0.089
REW9	0.759	0.712	0.031	-0.106	0.153	-0.003	-0.165	0.125	-0.004	0.028	**0.861**	0.357
REW10	0.795	0.602	0.090	0.001	0.212	0.149	-0.206	0.238	0.107	-0.034	**0.524**	-0.091

Extraction method: principal component analysis
Rotation method: promax with Kaiser-normalization

b) EFA Applied to Perceived Behavioral Control

In the context of the overall EFA (see Table 9), five of the six items relating to perceived behavioral control loaded highest on the same component. The item loading highest on a separate component (PBC6) was also conspicuous when the EFA was conducted separately for the construct (see Table 10). The communality of item PBC6 as well as the component loading did not meet the requirements, which suggested this item should be eliminated (cf. Hair et al. 2008, pp. 119f.; Weiber and Mühlhaus 2010, p. 109). This decision was further supported by scrutiny of its indicator loading (see Appendix C) and the impact of its exclusion on the construct's AVE and composite reliability (cf. Hair et al. 2014, p. 104). The indicator loading of item PBC6 was below the threshold value of 0.7 and the exclusion increased the construct's AVE, composite reliability, and even the Cronbach's alpha (cf. Appendix C and Table 14). In the end, the EFA did not support the claim perceived behavioral control comprises two independent variables, namely perceived self-efficacy and perceived controllability (see chapter 3.1.3).

c) *EFA Applied to Intention*

The overall EFA (see Table 9) confirmed it was worthwhile to distinguish between the intention to exchange documented knowledge in OI-projects and the intention to share undocumented knowledge with external partners. All intention-related items showed their highest loading on one of two components. I1 and I2 were supposed to measure intention to exchange documented knowledge and loaded on one component. I3, I4 and I5 were intended to measure intention to exchange undocumented knowledge and loaded on the other component. Thus, modeling intention as a second-order construct was again supported by the data (see chapter 4.4.1.4). When the EFA was conducted separately for the construct (see Table 10), all requirements were met, i.e., no item had to be removed.

d) *EFA Applied to Enjoyment in Helping*

During the overall EFA (see Table 9), the construct relating to enjoyment in helping was correctly identified, i.e., all items intended to measure enjoyment in helping showed the highest loading on the same component. Furthermore, all requirements were met when the EFA was conducted separately for the construct (see Table 10). Therefore, no adaptation was necessary. This decision was further supported after scrutinizing its indicator loading (see Appendix C and cf. Hair et al. 2014, p. 104).

e) *EFA Applied to Sense of Self-Worth*

The sense of self-worth construct was correctly identified in the overall EFA (see Table 9), i.e., all items intended to measure sense of self-worth showed the highest loading on the same component. When the EFA was conducted separately for the construct (see Table 10), the communality of item SW2 was insufficiently high, which suggested this item should be eliminated (cf. Hair et al. 2008, pp. 119f.; Weiber and Mühlhaus 2010, p. 109). This was further supported by the scrutiny of its indicator loading (see Appendix C) and the impact of its exclusion on the construct's AVE and composite reliability (cf. Hair et al. 2014, p. 104). The indicator loading of item SW2 was below the threshold value of 0.7 and the exclusion would improve the AVE considerably. Although another item (SW3) had a lower indicator loading than item SW2, the elimination of SW2 was preferable because its exclusion showed a stronger positive effect on AVE and the composite reliability. By excluding SW2, the AVE rose above the threshold of 0.5 and the composite reliability was slightly improved (cf. Appendix C and Table 14).

Table 10: MSA, Communalities and Pattern Matrix – All Reflective Constructs (Separately)

Construct	Item	MSA	Communality	Component Loading > 0.5	Initial Eigenvalue	Variance Extracted
Critical Value		> 0.5	> 0.4			> 50 %
Attitude	A1	0.818	0.394	0.628	2.338	46.768 %
	A2	0.804	0.425	0.652		
	A3	0.761	0.541	0.735		
	A4	0.799	0.459	0.677		
	A5	0.767	0.519	0.721		
Perceived Behavioral Control	PBC1	0.823	0.552	0.743	3.428	57.134 %
	PBC2	0.792	0.598	0.774		
	PBC3	0.888	0.607	0.779		
	PBC4	0.700	0.738	0.859		
	PBC5	0.706	0.699	0.836		
	PBC6	0.946	0.234	0.483		
Intention	I1	0.751	0.418	0.647	2.831	56.624 %
	I2	0.798	0.520	0.721		
	I3	0.850	0.625	0.791		
	I4	0.731	0.599	0.774		
	I5	0.749	0.668	0.818		
Enjoyment in Helping	JOY1	0.652	0.639	0.799	2.196	73.207 %
	JOY2	0.565	0.866	0.931		
	JOY3	0.622	0.691	0.831		
Sense of Self-Worth	SW1	0.834	0.462	0.680	2.528	50.557 %
	SW2	0.792	0.368	0.607		
	SW3	0.720	0.492	0.701		
	SW4	0.725	0.677	0.823		
	SW5	0.775	0.528	0.727		
Reciprocity	RP1	0.751	0.382	0.618	2.403	60.069 %
	RP2	0.659	0.638	0.799		
	RP3	0.643	0.751	0.866		
	RP4	0.742	0.632	0.795		

Extraction method: principal component analysis
Rotation method: promax with Kaiser-normalization

f) EFA Applied to Reciprocity

In the context of the overall EFA (see Table 9), three of the four reciprocity-related items loaded highest on the same component. The item loading highest on a separate component (RP1) was also conspicuous when the EFA was conducted separately for the construct (see Table 10). The communality of item RP1 did not meet the requirements, which suggested this item should be eliminated (cf. Hair et al. 2008, pp. 119f.; Weiber and Mühlhaus 2010, p. 109).

g) EFA Applied to Rewards

The overall EFA (see Table 9) revealed that the ten items relating to rewards measure multiple factors, since the highest loadings of the ten items were distributed over three components. This assumption was confirmed by the reward-specific EFA. As presented in Table 11, two independent components were extracted: "reward A" and "reward B". Furthermore, the results of the reward-specific EFA showed that item REW5, which had been the only item with a high loading on the third component in the overall EFA, did not meet the requirements regarding communality, suggesting it should be excluded from further analyses (cf. Hair et al. 2008, pp. 119f.; Weiber and Mühlhaus 2010, p. 109). This was further supported by scrutinizing its indicator loading (see Appendix C) and the impact of its exclusion on the construct's AVE and composite reliability (cf. Hair et al. 2014, p. 104). The indicator loading of REW5 was below the threshold value of 0.7 and the exclusion would improve the AVE of construct reward A considerably. In addition, the examination of the other reward items also showed the indicator loading of item REW9 was below the threshold value of 0.7 and its exclusion would improve the AVE of the reward B construct considerably. Therefore, both items were excluded from further analysis, resulting in a considerable improvement of both constructs' AVE and a slight increase in the composite reliability of reward A and reward B. Even the Cronbach's alpha of both constructs was slightly improved (cf. Appendix C and Table 14).

Table 11: MSA, Communalities and Pattern Matrix – Reward Construct

	MSA	Communality	Component 1	Component 2
Critical Value	> 0.5	> 0.4	Loading > 0.5	
REW1	0.807	0.427	**0.632**	0.081
REW2	0.830	0.792	**0.906**	-0.103
REW3	0.783	0.846	**0.935**	-0.098
REW4	0.831	0.828	**0.921**	-0.063
REW5	0.766	0.379	**0.598**	0.068
REW6	0.919	0.543	**0.612**	0.302
REW7	0.661	0.530	0.021	**0.723**
REW8	0.617	0.801	-0.055	**0.905**
REW9	0.686	0.421	0.174	**0.589**
REW10	0.766	0.452	-0.079	**0.685**
Initial Eigenvalue			3.982	2.038
Rotation Sums of Squared Loadings†			3.835	2.468
Cumulative Variance			60.204 %	

Extraction method: principal component analysis
Rotation method: promax with Kaiser-normalization
† When components are correlated, sums of squared loadings cannot be added to obtain a total variance.

h) *EFA Applied to Control Variable "Big Five"*

The construct-specific EFA was conducted for the control variable related to the "Big Five" personality traits. As shown in Table 12, the ten items, which were intended to measure five dimensions, loaded only highly on three components. Items BF1 and BF1r, which were put in place to measure "extraversion", and items BF5 and BF5r, which were intended to measure "openness to experience", both loaded highly on the same component suggesting these two personality traits are interrelated.[152] Since item BF5r did not meet the requirements regarding communality, it was eliminated (cf. Hair et al. 2008, pp. 119f.; Weiber and Mühlhaus 2010, p. 109). As a consequence, BF5 was removed in order to exclude all measures related to openness to experience. The items supposed to measure "agreeableness" (BF2 and BF2r) and "conscientiousness" (BF3 and BF3r) loaded highest on two different components as intended. The two items relating to "emotional stability" (BF4 and BF4r) did not load highly on one construct and so were excluded. In the end, only items measuring three dimensions (extraversion, agreeableness and conscientiousness) remained.

Table 12: MSA, Communalities and Pattern Matrix – Big Five Construct

	MSA	Communality	Component 1	Component 2	Component 3
Critical Value	> 0.5	> 0.4		Loading > 0.5	
BF1	0.561	0.735	**0.780**	0.240	-0.050
BF1r	0.568	0.579	**-0.641**	-0.293	0.083
BF2	0.667	0.602	0.197	**0.747**	0.106
BF2r	0.638	0.458	0.230	**-0.706**	-0.260
BF3	0.516	0.571	-0.077	0.221	**0.792**
BF3r	0.524	0.544	-0.043	-0.115	**-0.765**
BF4	0.620	0.618	-0.056	**0.656**	-0.284
BF4r	0.628	0.530	0.108	-0.193	**0.643**
BF5	0.591	0.484	**0.650**	-0.242	0.178
BF5r	0.677	0.376	**-0.592**	0.267	-0.014
Initial Eigenvalue			2.32	1.875	1.302
Rotation Sums of Squared Loadings†			1.939	1.986	1.923
Cumulative Variance				54.972 %	

Extraction method: principal component analysis
Rotation method: promax with Kaiser-normalization
† When components are correlated, sums of squared loadings cannot be added to obtain a total variance.

To sum up the EFA results, six items relating to model-relevant constructs and four items relating to the "Big Five" were excluded during the EFA. The elimination of items affected the following constructs (items): Attitude (A1); perceived behavioral control (PBC6); sense of self-worth (SW2); reciprocity (RP1); reward A (REW5); reward B (REW9) and "Big Five" (BF4, BF4r, BF5, BF5r). After excluding these items, a second overall EFA was employed. All reflective constructs were correctly identified, i.e., the intended constructs derived from theoretical considerations were all properly replicated, which indicates an improvement of the reflective measures (see Appendix D). Furthermore, the ITC of the remaining items and the IIC of the reflective constructs were acceptable, as shown in Table 13.

---

[152] The "r" in the item name indicates that it is a reversed-coded item.

Table 13: EFA Results, ITC, and IIC after Item Exclusion

Construct	Item	MSA	Commu-nality	Component Loading	Corrected Item-Total-Correlation	Inter-Item-Correlation	Variance Extracted
Critical Value		> 0.5	> 0.4	> 0.5	ITC ≥ 0.3	IIC ≥ 0.3	> 50 %
Attitude	A2	0.758	0.473	0.688	0.440	0.355	51.686 %
	A3	0.706	0.572	0.756	0.506		
	A4	0.759	0.478	0.691	0.444		
	A5	0.713	0.544	0.738	0.480		
Perceived Behavioral Control	PBC1	0.807	0.556	0.746	0.633	0.561	65.012 %
	PBC2	0.775	0.600	0.775	0.672		
	PBC3	0.880	0.619	0.786	0.657		
	PBC4	0.688	0.759	0.871	0.739		
	PBC5	0.694	0.716	0.846	0.700		
Intention	I1	0.751	0.418	0.647	0.490	0.455	56.624%
	I2	0.798	0.520	0.721	0.580		
	I3	0.850	0.625	0.791	0.633		
	I4	0.731	0.599	0.774	0.588		
	I5	0.749	0.668	0.818	0.648		
Enjoyment in Helping	JOY1	0.652	0.639	0.799	0.573	0.594	73.207 %
	JOY2	0.565	0.866	0.931	0.812		
	JOY3	0.622	0.691	0.831	0.630		
Sense of Self-Worth	SW1	0.803	0.507	0.712	0.495	0.420	56.736 %
	SW3	0.704	0.570	0.755	0.538		
	SW4	0.677	0.694	0.833	0.644		
	SW5	0.730	0.498	0.706	0.480		
Reciprocity	RP2	0.635	0.731	0.855	0.651	0.566	71.342 %
	RP3	0.598	0.822	0.907	0.755		
	RP4	0.757	0.587	0.766	0.530		
Reward A	REW1	0.903	0.427	0.654	0.527	0.586	67.972 %
	REW2	0.866	0.823	0.907	0.800		
	REW3	0.775	0.837	0.915	0.800		
	REW4	0.798	0.839	0.916	0.804		
	REW6	0.915	0.472	0.687	0.560		
Reward B	REW7	0.559	0.645	0.803	0.490	0.446	63.642 %
	REW8	0.538	0.806	0.898	0.696		
	REW10	0.620	0.459	0.677	0.373		

6.4.1.2 Confirmatory Factor Analysis (CFA) – A Second Generation Method

After the EFA was conducted, the measurement model of the reflective constructs with its remaining items had to be evaluated (see Netemeyer et al. 2003). In order to do this evaluation indicator reliability, internal consistency reliability, convergent validity, and discriminant validity all had to be checked (cf. Hair et al. 2011, pp. 145ff.; Hair et al. 2012, pp. 423f; Henseler et al. 2009, pp. 298ff.). This was done by applying a CFA – a second-generation method to check for reliability and validity of reflective measures (see Fornell 1982; Fornell 1987).[153] The results are explained in the following.

---

[153] As suggested by Hair et al. 2012, p. 429, I used the following PLS algorithm settings: path weighting scheme; data metric: mean 0, var 1; maximum iterations: 300; abort criterium: 10^{-5}; initial weights: 1

a) Indicator Reliability

The reliability of an indicator describes how much of the indicator's variance is explained by the respective latent variable (cf. MacKenzie et al. 2011, pp. 314f.). Generally, it is assessed by considering the absolute standardized indicator loadings (cf. Hair et al. 2012, p. 424; Schweisfurth 2013, p. 97). In the case of second-order constructs (e.g., the intention construct), the relation between the higher-order and lower-order constructs, i.e., the path coefficients between the second-order construct (intention) and the first-order constructs (intention_doc and intention_undoc) indicate the reliability of indicators (cf. Becker et al. 2012, p. 378; Wetzels et al. 2009, pp. 187f.).[154] To be considered acceptable, standardized indicator loadings and path coefficients should exceed 0.7 (cf. Henseler et al. 2009, p. 300; Hair et al. 2014, p. 104; Hulland 1999, p. 198). Despite this cut-off value, Hair et al. (2011, p. 145) suggest the rigorous elimination of items only where the indicator loading is below 0.4, i.e., when loadings range between 0.4 and 0.7 the decision to remove items should be made on a case-by-case basis.

As indicated in Table 14, most of the remaining items exceeded the critical value of 0.7 and, thus, were considered reliable. However, two items only possessed indicator loadings of 0.643 and 0.648, respectively. In order to decide on these items, I examined the composite reliability and AVE of both affected constructs (attitude and sense of self-worth) and found that for both constructs the threshold for composite reliability and AVE was already met.

Consequently, the elimination of the two items would not increase the measures above the cut-off value, since the thresholds had already been exceeded. This suggested the two items should be retained (cf. Hair et al. 2011, p. 145; Hair et al. 2014, p. 104). In addition, I conducted bootstrapping (cf. Chin 1998b, p. 320; Henseler et al. 2009, pp. 305ff.) with 133 cases and 8,000 samples to evaluate the significance of each item.[155] Both items were significant on a 0.1% significance level (see Table 14), which further strengthened the decision to keep the two items.

b) Internal Consistency Reliability

Internal consistency reliability refers to correlations among items (cf. Nunnally and Bernstein 1994, pp. 251f.) and can be assessed by means of two different criteria (cf. Weiber and Mühlhaus 2010, p. 122). The most prominent reliability coefficient is Cronbach's alpha (MacKenzie et al. 2011, p. 314; Hair et al. 2012, p. 424), which represents the average

---

[154] In Table 14, path 1 represents the relation between the second-order construct intention and the first-order construct intention_doc. Path 2 represents the relation between intention and intention_undoc.
[155] According to Hair et al. 2011, p. 145, the bootstrapping should be based on at least 5,000 samples and on the number of cases in the original sample (in this study N=133). In this study, bootstrapping was conducted with 133 cases and 8,000 samples (unless noted). The following critical t-values for a two-tailed test were considered satisfactory: 1.96 (p < 0.05: significance level = 5%), 2.58 (p < 0.01: significance level = 1%), and 3.29 (p < 0.001: significance level = 0.1%), i.e., a significance level of 10% (t-value: 1.65) was not deemed sufficient.

correlation between all items belonging to a specific construct (see Cortina 1993; Cronbach 1951).[156] Theoretically, Cronbach's alpha can range from 0 to 1. Nunnally and Bernstein (1994, pp. 264f.) introduced the widely accepted cut-off value of 0.7. However, according to Peterson (1994, pp. 388f.), Cronbach's alpha should not exceed 0.9. Table 14 shows that almost all constructs met the desired range. Two constructs were slightly below the cut-off value (attitude: 0.688; intention_doc: 0.681). A second criterion for the evaluation of internal consistency reliability is the Dillon-Goldstein's ρ (composite reliability) (cf. Henseler et al. 2009, p. 300). In contrast to Cronbach's alpha, Dillon-Goldstein's ρ does not assume equal reliability for all items, which fits better with the PLS algorithm (cf. Hair et al. 2012, p. 424). Therefore, composite reliability is the superior assessment of internal consistency if PLS is employed (cf. Chin 1998b, p. 320; Hair et al. 2012, p. 424). The critical value for Dillon-Goldstein's ρ is 0.7 (cf. Hair et al. 2011, p. 145; Henseler et al. 2012, p. 269), which had been achieved by all constructs (see Table 14). In summary, almost all constructs met both criteria for internal consistency reliability. Attitude and intention_doc slightly missed the critical value for Cronbach's alpha but clearly exceeded it for Dillon-Goldstein's ρ.

c) Convergent Validity

Convergent validity shows to what extent a set of items represent their theoretically intended construct (cf. Henseler et al. 2009, p. 299). The assessment is carried out by examining the AVE, which denotes how much of the indicator's variance is explained by the theoretical construct (cf. Hair et al. 2011, p. 146). In the case of second-order constructs, the AVE equals the average of first-order constructs' squared multiple correlations[157] (cf. MacKenzie et al. 2011, p. 313). In order to fulfill the requirement for convergent validity, it must exceed a critical value of 0.5 (cf. Bagozzi and Yi 1988, p. 80; Fornell and Larcker 1981, p. 46), which means that at least 50% of the indicator's variance is explained by the construct. As indicated in Table 14, all constructs achieved such a cut-off value.

d) Discriminant Validity

Discriminant validity goes one step further than convergent validity. It not only considers the relationship between a set of items and their theoretically intended construct, but also mandates that items should be stronger related to their intended construct than to any other construct in the study (cf. Henseler et al. 2009, p. 299).

In order to evaluate discriminant validity, the Fornell-Larcker-criterion and cross-loadings have to be examined (cf. Hair et al. 2011, p. 145; Henseler et al. 2012, p. 269). The Fornell-Larcker-criterion postulates that the squared root AVE of a construct should be higher than the correlations between this construct and all other constructs in the study (cf. Fornell and

---

[156] In the case of second-order constructs, Cronbach's alpha is only calculated for the first-order constructs (see Wetzels et al. 2009).
[157] Squared multiple correlation is denoted R^2 (see chapter 6.5).

Larcker 1981, p. 46). Table 15 displays the correlations between all reflective constructs – the bold numbers on the diagonal indicate the squared root of the constructs' AVE.

Table 14: Indicator and Internal Consistency Reliability, Convergent Validity

		INDICATOR RELIABILITY		INTERNAL CONSISTENCY RELIABILITY		CONVERGENT VALIDITY
		Standardized Indicator Loading λ	T-Value	Dillon-Goldstein's ρ	Standardized Cronbach's α	Average Variance Extracted
Critical Value		λ ≥ 0.7	≥ 1.96: p<0.05 ≥ 2.58: p<0.01 ≥ 3.29: p<0.001	ρ ≥ 0.7	0.7 ≤ α ≤ 0.9	AVE ≥ 0.5
Construct	Item					
Attitude	A2	0.715	10.588	0.809	0.688	0.515
	A3	0.767	16.022			
	A4	0.643	7.624			
	A5	0.740	14.083			
Perceived Behavioral Control	PBC1	0.727	14.104	0.902	0.865	0.649
	PBC2	0.777	18.861			
	PBC3	0.811	22.955			
	PBC4	0.866	23.089			
	PBC5	0.840	20.270			
Intention (2nd order)	Path_1	0.802	20.663	0.867	n.a.	0.748
	Path_2	0.923	66.287			
Intention_doc (1st order)	I1	0.857	26.726	0.862	0.681	0.758
	I2	0.884	46.036			
Intention_undoc (1st order)	I3	0.816	21.181	0.888	0.811	0.727
	I4	0.861	27.801			
	I5	0.879	37.019			
Enjoyment in Helping	JOY1	0.857	25.183	0.887	0.814	0.724
	JOY2	0.928	52.172			
	JOY3	0.760	9.633			
Sense of Self-Worth	SW1	0.740	12.580	0.833	0.744	0.557
	SW3	0.648	7.354			
	SW4	0.804	14.570			
	SW5	0.782	12.861			
Reciprocity	RP2	0.820	7.647	0.879	0.797	0.708
	RP3	0.878	9.673			
	RP4	0.826	10.300			
Reward A	REW1	0.761	3.448	0.906	0.876	0.659
	REW2	0.865	3.953			
	REW3	0.835	3.590			
	REW4	0.845	3.781			
	REW6	0.746	3.437			
Reward B	REW7	0.803	16.250	0.838	0.707	0.635
	REW8	0.874	24.046			
	REW10	0.704	8.940			

Bootstrapping conducted with 133 cases and 8,000 samples

The examination of Table 15 reveals that all constructs met the Fornell-Larcker-criterion. With regard to cross-loadings, discriminant validity is given if each item loads highest on its assigned constructs and not unintentionally on other latent variables (cf. Chin 1998b,

p. 321). By observing Table 16, it becomes evident this criterion is also fulfilled in all cases. Thus, both criteria for discriminant validity were fully met.

Table 15: Correlations and Discriminant Validity

		1	2	3	4	5	6	7	8	9
1	Attitude	**0.718**								
2	Perceived B. Control	0.280	**0.806**							
3	Intention_doc	0.280	0.366	**0.871**						
4	Intention_undoc	0.380	0.443	0.511	**0.852**					
5	Enjoyment in Helping	0.477	0.122	0.224	0.375	**0.851**				
6	Self-Worth	0.382	0.077	0.252	0.356	0.362	**0.746**			
7	Reciprocity	0.247	0.217	0.249	0.275	0.325	0.269	**0.842**		
8	Reward A	0.155	-0.097	0.015	0.001	0.246	0.124	0.277	**0.812**	
9	Reward B	0.406	0.445	0.295	0.444	0.372	0.389	0.371	0.224	**0.797**

Note: Bold numbers on the diagonal illustrate the squared root of the AVE.

Table 16: Cross-Loadings

	Attitude	Perceived B. Control	Intention _doc	Intention _undoc	Enjoyment in Helping	Self-Worth	Reciprocity	Reward A	Reward B
A2	**0.715**	0.150	0.087	0.310	0.411	0.266	0.161	0.132	0.353
A3	**0.767**	0.279	0.207	0.317	0.308	0.348	0.177	0.150	0.370
A4	**0.643**	0.221	0.241	0.123	0.288	0.251	0.222	0.034	0.159
A5	**0.740**	0.161	0.289	0.303	0.354	0.229	0.168	0.107	0.248
PBC1	0.196	**0.727**	0.185	0.297	0.044	-0.015	0.197	-0.087	0.355
PBC2	0.306	**0.777**	0.246	0.385	0.076	0.053	0.079	-0.017	0.325
PBC3	0.175	**0.811**	0.345	0.413	0.119	0.123	0.204	-0.009	0.330
PBC4	0.228	**0.866**	0.311	0.340	0.143	0.053	0.194	-0.136	0.398
PBC5	0.228	**0.840**	0.356	0.332	0.095	0.067	0.203	-0.156	0.394
I1	0.168	0.354	**0.857**	0.400	0.118	0.143	0.255	0.023	0.252
I2	0.313	0.287	**0.884**	0.486	0.266	0.289	0.183	0.004	0.262
I3	0.309	0.411	0.490	**0.816**	0.341	0.299	0.244	0.065	0.329
I4	0.344	0.350	0.369	**0.861**	0.295	0.334	0.224	-0.025	0.411
I5	0.320	0.371	0.445	**0.879**	0.320	0.280	0.234	-0.039	0.395
JOY1	0.458	0.196	0.227	0.337	**0.857**	0.341	0.230	0.151	0.350
JOY2	0.454	0.072	0.221	0.335	**0.928**	0.325	0.299	0.239	0.345
JOY3	0.248	-0.003	0.082	0.278	**0.760**	0.238	0.341	0.280	0.227
SW1	0.304	0.157	0.155	0.373	0.351	**0.740**	0.204	0.071	0.400
SW3	0.149	-0.191	0.108	0.165	0.202	**0.648**	0.047	0.121	0.131
SW4	0.277	-0.061	0.136	0.197	0.293	**0.804**	0.209	0.132	0.156
SW5	0.347	0.173	0.301	0.284	0.223	**0.782**	0.266	0.072	0.382
RP2	0.187	0.154	0.133	0.091	0.225	0.174	**0.820**	0.310	0.238
RP3	0.185	0.188	0.240	0.197	0.247	0.200	**0.878**	0.213	0.300
RP4	0.242	0.200	0.245	0.364	0.329	0.284	**0.826**	0.187	0.377
REW1	0.159	-0.093	0.119	0.061	0.302	0.150	0.307	**0.761**	0.205
REW2	0.125	-0.181	-0.056	-0.065	0.121	0.141	0.154	**0.865**	0.079
REW3	0.061	-0.115	-0.077	-0.082	0.104	0.075	0.119	**0.835**	0.085
REW4	0.084	-0.082	-0.048	-0.062	0.066	0.127	0.159	**0.845**	0.111
REW6	0.135	0.055	0.025	0.067	0.268	-0.004	0.268	**0.747**	0.326
REW7	0.344	0.337	0.141	0.248	0.300	0.199	0.213	0.221	**0.803**
REW8	0.312	0.407	0.227	0.445	0.427	0.282	0.387	0.191	**0.874**
REW10	0.310	0.317	0.346	0.376	0.160	0.456	0.293	0.117	**0.705**

6.4.2 Formative Constructs

Due to the underlying differences between reflective and formative measurement models, concepts such as internal consistency reliability and convergent validity are not meaningful in cases of formative measures, i.e., it is not possible to apply the quality criteria for reflective scales (e.g., composite reliability, AVE) to formative constructs (cf. Bagozzi 1994, p. 333; Hair et al. 2011, p. 146; Hair et al. 2012, p. 424). Nevertheless, the quality of formative measures must also be evaluated. As formative constructs are characterized by items causing the construct, it is essential all relevant aspects are considered, i.e., that the set of items is "complete". Therefore, the qualitative evaluation of formative constructs is particularly important. Content and face validity can be considered as a first quality criterion. The indicators' weights and loadings (including the level of significance) are a second reference for the quality of formative measures, as it indicates whether every item does indeed contribute to the construct (cf. Henseler et al. 2009, p. 302). A third criterion is related to redundancy of items, which is assessed by the level of multicollinearity among the formative items (cf. Hair et al. 2011, p. 146). As indicated in chapter 4.4.1, only the subjective norm construct is operationalized with formative measures. To assess the quality of this construct, the three introduced criteria for formative measurement models are discussed in the following.

6.4.2.1 Content and Face Validity

Even though content validity and face validity are quite similar and often used interchangeably, they do not refer to exactly the same phenomena (cf. Netemeyer et al. 2003, pp. 12f.). According to Nunnally and Bernstein (1994, pp. 101ff.), content validity refers to the ability of an item-block to measure the intended construct and results from a plan containing content and construction of the items, which is made before the questionnaire is actually developed. Face validity, on the other hand, is compiled after the questionnaire's construction and refers to the extent to which people feel the items measure the intended construct (cf. Bryman 2008, p. 152; Nunnally and Bernstein 1994, p. 110). In this study, content validity of the subjective norm construct is ensured because I applied measures that were carefully developed and used many times previously by other researchers (see chapter 4.4.1.2). The study of Karahanna et al. (1999) provides further evidence that I considered the most important sources for social pressure within a professional context. Face validity is proven because the items related to subjective norm were checked from academic and company representatives during the pre-test (see chapter 4.4.2). Consequently, both qualitative evaluation criteria were met.

6.4.2.2 Indicator Weights and Loadings

The indicator weight and indicator loading of an item provide information about its importance; where indicator weights represent the relative importance and loadings the absolute importance of items (cf. Hair et al. 2011, pp. 145f.). In addition, it is advisable to perform bootstrapping in order to evaluate the item's significance. Table 17 indicates the importance and significance of all three items related to subjective norm.[158] The social pressure caused by the CEO (SN1) and colleagues (SN3) was highly significant and showed high relative as well as absolute importance. Consequently, both items contributed massively to the subjective norm construct. Social pressure caused by the immediate supervisor (SN 2) had a highly significant absolute importance; but the relative importance was neither very strong nor significant. Consequently, the contribution of this item is smaller but still valid (cf. Hair et al. 2011, pp. 145f.).

Table 17: Evaluation of Formative Measures of Subjective Norm

	Indicator Weight	Indicator Weight's T-Value	Indicator Loading	Indicator Loading's T-Value	Tolerance	Variance Inflation Factor
Critical Value		≥ 1.96: p<0.05 ≥ 2.58: p<0.01 ≥ 3.29: p<0.001		≥ 1.96: p<0.05 ≥ 2.58: p<0.01 ≥ 3.29: p<0.001	> 0.2	VIF < 5
SN1 (CEO)	0.542	3.910	0.860	13.382	0.618	1.619
SN2 (supervisor)	0.146	0.873	0.698	5.890	0.614	1.630
SN3 (colleagues)	0.522	3.833	0.829	11.561	0.746	1.340

Bootstrapping conducted with 133 cases and 8,000 samples

6.4.2.3 Multicollinearity

The degree of multicollinearity provides information about correlations between the formative items and can be assessed by way of two measures: tolerance and variance inflation factor (VIF) (cf. Hair et al. 2014, p. 154; Henseler et al. 2009, p. 302). Both measures are interrelated, since the VIF is the tolerance's inverse (cf. Hair et al. 2008, p. 201). Tolerance values can range from 0 to 1, where 1 represents the absence of multicollinearity (cf. Hair et al. 2008, p. 201). Thus, tolerance should be as high as possible and the VIF should be as low as possible (the lowest possible value for VIF is 1). A widely accepted rule of thumb says VIF should not exceed 5 (cf. Hair et al. 2011, p. 145; Hair et al. 2012, p. 430), which implies that tolerance values above 0.2 are acceptable. Table 17 shows all tolerance values were far above the cut-off value and so the VIF of all the items can be considered adequate. Therefore, I have concluded that multicollinearity is not a great issue for this study.

---

[158] Each item is related to the social pressure caused by a specific person, and group respectively (CEO, immediate supervisor, and colleagues).

6.5 Structural Model

After the assessment of the measurement model (outer model) had proven satisfactory, the structural model (inner model) was evaluated with respect to quality and hypothesized relationships. In contrast to covariance-based structural equation models, PLS does not provide a global measure to estimate the structural model's fit, i.e., the overall quality (Henseler et al. 2012, p. 267).[159] Therefore, the evaluation of the inner model in PLS is based on several single criteria, which consider different aspects of the structural model and allow assessment of the model (cf. Weiber and Mühlhaus 2010, pp. 254ff.) The most common criteria are path coefficients and their significance, the explained variance and predictive relevance (cf. Chin 1998b, pp. 316ff.; Hair et al. 2012, pp. 426f.; Henseler et al. 2012, p. 271).

"The individual path coefficients of the PLS structural model can be interpreted as standardized beta coefficients of ordinary least squares regressions." (Hair et al. 2011, p. 147) According to Chin (1998a, p. xiii), a path coefficient[160] may be considered meaningful if a critical value of 0.2 is exceeded. Analog to the significance of items, path coefficient's significance were evaluated by conducting bootstrapping (133 cases and 8,000 samples) and the following critical t-values for a two-tailed test were considered satisfactory in this study:

- ✓ 1.96 ($p < 0.05$: significance level = 5%)
- ✓ 2.58 ($p < 0.01$: significance level = 1%)
- ✓ 3.29 ($p < 0.001$: significance level = 0.1%)

The variance explained is a further fundamental criterion for the assessment of the inner model and is symbolized by R^2. The R^2 value of an endogenous or dependent latent variable provides information on how much variance is explained by the connected exogenous or independent latent variables (cf. Hair et al. 2012, p. 426). It can range from 0 to 1, where $R^2 = 0$ can be interpreted as explaining no variance and $R^2 = 1$ would mean that 100% of the variance is explained by the exogenous latent variables. The requirements regarding the level of R^2 depend on the research field but in disciplines such as consumer behavior even values of 0.2 are considered high (cf. Hair et al. 2011, p. 147). To understand how strongly a dependent latent variable is influenced by specific independent constructs, the effect size

---

[159] Tenenhaus et al. 2004 suggested a global goodness-of-fit (GoF) criterion for PLS, which is calculated "[...] *as the geometric mean of the average communality and the average R^2* [...]" (Tenenhaus et al. 2005, p. 173). However, Henseler and Sarstedt 2013 argued that the GoF has several weaknesses (it is only appropriate for reflective and multi-item constructs and ignores potential overparameterization) and demonstrated the inappropriateness of the GoF for a global model validation by conducting an empirical investigation.

[160] In the following, b will be used as symbol for path coefficients.

f^2 needs to be consulted for each specific independent construct and can be calculated as follows (cf. Chin 1998b, pp. 316f.):

$$f^2 = \frac{R^2_{included} - R^2_{excluded}}{1 - R^2_{included}}$$

$R^2_{included}$ = R^2 of dependent variable if the specific independent variable is used

$R^2_{excluded}$ = R^2 of dependent variable if the specific independent variable is omitted

With respect to the interpretation of effect size f^2, values greater than 0.02, 0.15, and 0.35 can be considered as small, medium, and large effects respectively (cf. Chin 1998b, p. 317 based on Cohen 1988, p. 355).

The last quality criterion for the evaluation of the structural model is the predictive relevance, which is symbolized by Q^2. It can be traced back to Stone (1974) and Geisser (1974) and is therefore also known as Stone-Geisser criterion. It assesses the model's ability to adequately predict each indicator of the dependent variable (cf. Hair et al. 2011, p. 147; Henseler et al. 2009, p. 305). In PLS, Q^2 is calculated by applying blindfolding (cf. Chin 1998b, p. 317).[161] If Q^2 is greater than zero, the inner model has predictive relevance for the dependent construct (cf. Chin 1998b, p. 318; Hair et al. 2011, p. 145). The relative predictive relevance q^2 of a specific independent construct on the dependent latent variable has to be consulted and can be calculated as follows (cf. Chin 1998b, p. 318):

$$q^2 = \frac{Q^2_{included} - Q^2_{excluded}}{1 - Q^2_{included}}$$

$Q^2_{included}$ = Q^2 of dependent variable if the specific independent variable is used

$Q^2_{excluded}$ = Q^2 of dependent variable if the specific independent variable is omitted

The interpretation of the relative predictive relevance q^2 is analogous to the interpretation of effect size f^2, i.e., values greater than 0.02, 0.15, and 0.35 can be considered small, medium, and large degrees of predictive relevance respectively (cf. Henseler et al. 2009, p. 305).

6.5.1 Evaluation of Structural Model and Hypotheses

In the following, I apply the quality criteria to my structural model and evaluate whether the data support the hypotheses derived from the literature in chapter 3.3. The results from the structural model are presented in Figure 25.

---

[161] To perform blindfolding in PLS, an Omission Distance (OD) has to be determined. This defines the data points to be omitted from the data set and subsequently predicted by using information from the remaining data set (for more details on the procedure cf. Chin 1998b, pp. 317f., Tenenhaus et al. 2005, pp. 174ff.). According to Chin 1998b, p. 318, the quotient of sample size (i.e., N = 133) and OD should not be equal to an integer and OD should be between five and ten. In this study, blindfolding was conducted with OD = 8.

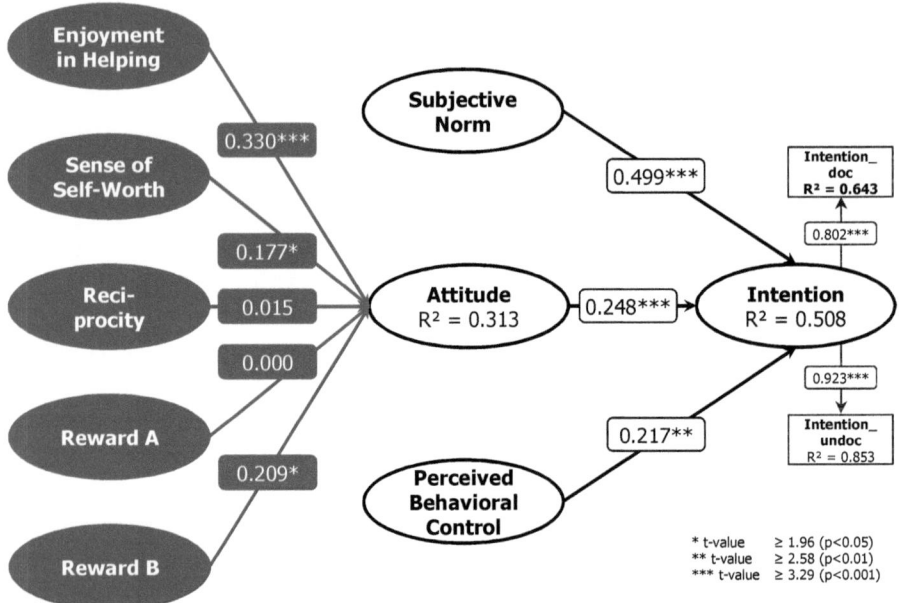

Figure 25: Results from PLS Analysis[162]

Considering all of the independent variables linked to the dependent attitude construct, enjoyment in helping had the strongest and most significant positive impact, with a path coefficient of 0.330 and significance level of 0.1% ($p < 0.001$). Reward B (i.e., the rewards mentioned by the pre-testers) also had a meaningful and significant positive influence on attitude ($b = 0.209$; $p < 0.05$). Sense of self-worth showed a significant relation but had only a moderately positive impact ($b = 0.177$; $p < 0.05$). Reciprocity and reward A were found to have no significant or meaningful influence on attitude. Considering all of the constructs linked to the dependent variable intention, subjective norm had by far the strongest and most significant positive impact ($b = 0.499$; $p < 0.001$). The link between subjective norm and intention was the strongest and most significant relationship in the entire structural model. Attitude was also found to have a meaningful and highly significant positive impact on intention ($b = 0.248$; $p < 0.001$), followed by perceived behavioral control, which also positively influenced intention on a meaningful and significant level ($b = 0.217$; $p < 0.01$).

The variance in the two dependent variables was explained to a substantial extent. The value of R^2 for attitude was 0.313, meaning that the model explained 31% of the variance in attitude. With respect to intention, a R^2 value of 0.508 could be reached, i.e., 51% of

---

[162] Author's illustration

intention's variance was explained by the model. As indicated in Table 18, all exogenous variables with a significant link to one of the two endogenous variables showed a noticeable effect size f^2 if the interpretation is based on the previously introduced cut-off values suggested by Chin (1998b, p. 317). With $f^2 = 0.437$, subjective norm had a large effect on intention. Attitude and perceived behavioral control had a small effect on intention ($f^2_{A-I} = 0.114$; $f^2_{PBC-I} = 0.079$). With respect to attitude being the endogenous variable, enjoyment in helping, sense of self-worth, and reward B each had a small impact on attitude ($f^2_{JOY-A} = 0.116$; $f^2_{SW-A} = 0.033$; $f^2_{REWB-A} = 0.047$). The variables with a non-significant relation to attitude (i.e., reciprocity and reward A) did not have any effect on the dependent variable.

Table 18: Evaluation of Structural Model

Endogenous Variable	$R^2_{incl.}$	$Q^2_{incl.}$	Exogenous Variable	Path Coefficient	T-Value	f^2	q^2
Critical Value		> 0		> 0.2	≥ 1.96: p<0.05 ≥ 2.58: p<0.01 ≥ 3.29: p<0.001	> 0.02: small effect/degree > 0.15: medium effect/degree > 0.35: large effect/degree	
Intention	0.508	0.286	Attitude	0.248	4.293	0.114	0.050
			Subjective Norm	0.499	9.423	0.437	0.168
			Perceived Behavioral Control	0.217	3.031	0.079	0.029
Attitude	0.313	0.140	Enjoyment in Helping	0.330	3.958	0.116	0.045
			Self-Worth	0.177	2.311	0.033	0.012
			Reciprocity	0.015	0.222	0.000	-0.001
			Reward A	0.000	0.006	0.000	-0.005
			Reward B	0.209	2.369	0.047	0.015

Blindfolding conducted with an OD of 8; bootstrapping conducted with 133 cases and 8,000 samples

With respect to predictive relevance, Table 18 shows that both exogenous variables had a Q^2 greater than zero ($Q^2_{Intention} = 0.286$; $Q^2_{Attitude} = 0.140$), which implies the model appropriately predicted both constructs. Following the interpretation suggested by Henseler et al. (2009, p. 305), subjective norm was found to have a medium degree of predictive relevance related to intention ($q^2_{SN-I} = 0.168$), while attitude and perceived behavioral control showed a small degree of predictive relevance ($q^2_{A-I} = 0.050$; $q^2_{PBC-I} = 0.029$). Regarding attitude as an exogenous variable, only enjoyment in helping had a small degree of predictive relevance ($q^2_{JOY-A} = 0.045$). Sense of self-worth, reciprocity, reward A, and reward B were found to have no meaningful degree of predictive relevance.

Table 19: Evaluation of Hypotheses

Hypothesis	Path Coefficient	T-Value	Significance	Support of Hypothesis
H1: Attitude → Intention (positive)	0.248	4.293	***	Supported
H2: Subjective Norm → Intention (positive)	0.499	9.423	***	Supported
H3: Perceived Behavioral Control → Intention (positive)	0.217	3.031	**	Supported
H4: Enjoyment in Helping → Attitude (positive)	0.330	3.958	***	Supported
H5: Sense of Self-Worth → Attitude (positive)	0.177	2.311	*	Supported
H6: Reciprocity → Attitude (positive)	0.015	0.222	n.s.	Not supported
H7a: Reward A → Attitude (negative)	0.000	0.006	n.s.	Not supported
H7b: Reward B → Attitude (positive)	0.209	2.369	*	Supported

T-value: n.s. < 1.96, * ≥ 1.96 (p < 0.05), ** ≥ 2.58 (p < 0.01), *** ≥ 3.29 (p < 0.001)

After assessing the structural model based on the criteria introduced in chapter 6.5, the results of this evaluation will now be used to evaluate the hypotheses. As shown in Table 19, all hypotheses with the exception of hypothesis 6 and 7a (H6, H7a) were supported by the data. A detailed discussion will follow in chapter 7.

6.5.2 Evaluation of Control Variables

For the assessment of the effect of control variables on the two endogenous variables and the relations between endogenous and exogenous constructs, I followed Kock et al. (2008, pp. 187ff.) and Kock (2011, pp. 3f.) and directly linked all control variables to attitude as well as intention.[163]

Table 20 contrasts the results of the structural model calculated with and without control variables. All links between the endogenous variable, intention, and its exogenous constructs (attitude, subjective norm, and perceived behavioral control) were significant when control variables were both included and excluded, i.e., the associations between intention and its three independent variables were significant regardless of the control variables (cf. Kock 2011, p. 4 for interpretation).

---

[163] Some of the control variables were not numerical originally and, thus, had to be recoded before they were included in the dataset (e.g., gender: male/female recoded into 0/1). The responses for the control variable "name of educational field" were so diverse that the answers could not be assigned to a reasonable number of categories without losing information. Therefore, this control variable was not recoded and so excluded in the dataset.

Table 20: Comparison of Structural Model with and without Control Variables

Endogenous Variable	Exogenous Variable	CONTROL VARIABLES NOT INCLUDED			CONTROL VARIABLES INCLUDED		
		R^2	Path Coefficient	T-Value	R^2	Path Coefficient	T-Value
Intention	Attitude	0.508	0.248***	4.293	0.587	0.284 ***	3.842
	Subjective Norm		0.499***	9.423		0.564 ***	8.307
	Perceived Behavioral Control		0.217**	3.031		0.212 *	2.423
Attitude	Enjoyment in Helping	0.313	0.330***	3.958	0.405	0.383 ***	4.210
	Self-Worth		0.177*	2.311		0.111	1.100
	Reciprocity		0.015	0.222		-0.019	0.202
	Reward A		0.000	0.006		0.023	0.244
	Reward B		0.209*	2.369		0.171	1.677

Bootstrapping conducted with 133 cases/8,000 samples (control variables not included) and 120 cases/ 5,000 samples (control variables included), respectively
T-value: * ≥ 1.96 (p < 0.05), ** ≥ 2.58 (p < 0.01), *** ≥ 3.29 (p < 0.001)

With respect to the endogenous attitude construct, only enjoyment in helping showed significant relationships with attitude, when control variables were included and excluded. Reciprocity and reward A neither had a significant link to attitude when the control variables were excluded nor when they were included. It could be concluded therefore that reciprocity and reward A were not significantly associated with attitude regardless of the control variables. However, the control variables had an effect on sense of self-worth and reward B, which were both significantly – although not strongly – linked to attitude when the control variables were excluded. After including the control variables, the path coefficient decreased and the relation to attitude became insignificant in both cases. The reduction of the path coefficients is due to the involvement of multiple endogenous variables and a very high number of control variables, which normally leads to an artificial reduction of path coefficients (cf. Kock 2011, p. 4). In summary, the control variables explained 7.9% of intention's variance and 9.2% of attitude's variance, which is acceptable due to the very high number of control variables (15 control variables were added to the structural model).[164]

---

[164] The variance explained by the control variables was calculated by subtracting the R^2 (including control variables) from the R^2 (without control variables), i.e., for intention: 0.587 – 0.508 = 0.079, which equals 7.9% of the explained variance.

Table 21: Evaluation of Structural Model with Control Variables Included

Endogenous Variable	R²	Exogenous Variable	Path Coefficient	T-Value	Significance
Intention	0.587	Attitude	0.284	3.842	***
		Subjective norm	0.564	8.307	***
		Perceived behavioral control	0.212	2.423	*
		OI-projects	0.056	0.828	n.s.
		Partner 1	0.005	0.056	n.s.
		Partner 2	-0.119	1.335	n.s.
		Partner 3	0.041	0.513	n.s.
		Partner 4	-0.079	1.072	n.s.
		Partner 5	0.021	0.274	n.s.
		Age	0.164	1.291	n.s.
		Gender	0.148	2.103	*
		Big5_agreeableness	-0.036	0.387	n.s.
		Big5_conscientiousness	-0.094	1.128	n.s.
		Big5_extraversion	0.025	0.250	n.s.
		Highest education	0.008	0.094	n.s.
		Tenure	-0.166	1.282	n.s.
		Location	-0.028	0.400	n.s.
		Company	0.223	3.096	**
Attitude	0.405	Enjoyment in helping	0.383	4.210	***
		Sense of self-worth	0.111	1.100	n.s.
		Reciprocity	-0.019	0.202	n.s.
		Reward A	0.023	0.244	n.s.
		Reward B	0.171	1.677	n.s.
		OI-projects	0.102	0.834	n.s.
		Partner 1	0.047	0.399	n.s.
		Partner 2	0.063	0.564	n.s.
		Partner 3	-0.018	0.194	n.s.
		Partner 4	-0.044	0.428	n.s.
		Partner 5	0.036	0.325	n.s.
		Age	0.032	0.195	n.s.
		Gender	-0.009	0.111	n.s.
		Big5_agreeableness	0.087	0.835	n.s.
		Big5_conscientiousness	0.095	0.964	n.s.
		Big5_extraversion	0.060	0.542	n.s.
		Highest education	-0.019	0.168	n.s.
		Tenure	0.159	0.888	n.s.
		Location	0.050	0.569	n.s.
		Company	-0.139	1.635	n.s.

Bootstrapping conducted with 120 cases and 5,000 samples
T-value: n.s. < 1.96, * ≥ 1.96 ($p < 0.05$), ** ≥ 2.58 ($p < 0.01$), *** ≥ 3.29 ($p < 0.001$)

Table 21 displays path coefficients, t-values, and levels of significance of all endogenous variables (including the 15 control variables). The majority of the control variables did not have a significant association with an exogenous variable. However, gender and respondents' company affiliation had a significant effect on attitude, even though only the path coefficient between company affiliation and attitude was above 0.2 and so could be considered as meaningful (cf. Chin 1998a, p. xiii).

6.6 Chapter Summary

The analysis of the survey responses showed that on average R&D employees worked on 4.7 OI-projects during the last three years. Most experience resulted from OI-projects with universities/research institutes followed by customers. Respondents indicated the greater likelihood of exchanging undocumented knowledge with external partners than sharing documented knowledge due to confidentiality, efficiency, and reasons flowing from the quality of information. Asked about requirements for knowledge exchange in OI-projects, most answers related to legal security (e.g., NDA, IP-rights), followed by a good rapport (e.g., mutual trust), common ground and fairness.

Based on the TPB, R&D employees' intention to exchange knowledge in OI-projects was expected to be determined by their attitude, subjective norm, and perceived behavioral control. Indeed, all three factors were found to positively influence R&D employees' intention to exchange their knowledge in OI-projects. However, the perceived social pressure (subjective norm) had by far the strongest and most significant impact. The influence of attitude and perceived behavioral control were comparatively strong.

Based on the literature review and the pre-test, enjoyment in helping, sense of self-worth, reciprocity and reward B were expected to positively influence employees' attitude toward knowledge exchange in OI-projects. Reward A, however, was expected to have a negative impact. Enjoyment in helping showed the strongest influence on attitude, followed by reward B and sense of self-worth. Reward A and reciprocity did not show any effect on employees' attitude.

7 Discussion

In this chapter, findings from the interviews and online survey (see chapter 5 and 6) are discussed along the lines of the three research questions outlined in chapter 2.3. The findings are compared with – and related to – prior research to form a holistic view of the research questions and to answer them. Furthermore, the follow-up group discussions and the literature are consulted to find an explanatory approach for those hypotheses that were not supported by the data. The first sub-chapter exposes R&D managers' interpretations of open innovation and discusses aspects that are – from an R&D point of view – especially important for knowledge exchange in OI-projects (RQ1). The second sub-chapter reveals which factors determine the intention of R&D employees to exchange knowledge with external partners in OI-projects (RQ2). The third sub-chapter examines which motivational factors can positively influence R&D employees' willingness to exchange their knowledge in OI-projects (RQ3).

7.1 RQ1: R&D Perspective on Open Innovation

The purpose of the first research question was to better understand the meaning of open innovation from an R&D perspective and the aspects considered to be especially important for knowledge exchange in OI-projects. To answer these questions, findings from the interviews and from the open-ended survey question regarding basic conditions for knowledge exchange in OI-projects were consulted (see chapter 5 and 6.3).

The interviews with R&D managers shows open innovation means for them – in the broadest sense – their company's ambition to open up the innovation process and to co-operate with external partners to accelerate their own innovation processes, i.e., each innovation project conducted with at least one external partner would be defined as open innovation. But aside from this broadly formulated definition, interviewees additionally emphasized and repeatedly highlighted two important characteristics of open innovation.

Firstly, the external partner had to bring expertise that was not, or only to an insufficient degree, available within the host company. The external know-how serves to complement internal knowledge and so contributes to the improvement and acceleration of the host company's innovation process.

Secondly, the interviewees understood open innovation as close, intensive, and systematic co-operation with a partner. Problems were not only explained to external partners and solutions demanded; all participating partners learned from each other through intensive and iterative knowledge exchange. This conflicts with OI-related concepts such as crowdsourcing. Even though the interviewees theoretically considered crowdsourcing to be open innovation, it was rarely, if ever, applied and they obviously differentiated between the open, anonymous, and worldwide call for proposals (i.e., crowdsourcing) and the above described form of open innovation (i.e., close collaboration with iterative exchange).

All in all, the R&D perspective is basically in line with the OI-literature. Both are similar in that they emphasize the importance of external knowledge for accelerating the internal innovation process (see Chesbrough 2003, 2006c). In accordance with Miotti and Sachwald (2003) and the resource-based perspective, complementary resources are considered to be a major reason for open innovation. Furthermore, the preference of companies for the coupled OI-process (i.e., inflows and outflows of knowledge) identified in the literature (see Lichtenthaler 2008; Schroll and Mild 2011; Vrande et al. 2009) was confirmed by the interviews. However, in some areas theory and practice run contrary to each other. The close co-operation with external partners, which was declared as central element of open innovation by the R&D managers, conflicts with OI-related concepts such as crowdsourcing (see Howe 2009) because further inquiries to an external partner and intensive co-operation are not feasible. Another point relates to the knowledge exchanged within OI-projects. In the literature, knowledge inflows and outflows are addressed only in a generic way, without reference to the types of knowledge being exchanged. The formulation given by the R&D managers, on the other hand, is more specific and limits the knowledge flow largely to complementary knowledge.

In order to identify aspects of particular importance to knowledge exchange in OI-projects, the interview findings and the results from the open-ended survey question were considered (see Figure 26). Basically, the interviews suggested that the processes of OI-projects and closed innovation projects do not significantly differ from each other. Both innovation project types follow a stage-gate process (see Cooper 1990, 1996). However, in line with the literature (see Slowinski and Sagal 2010; West and Bogers 2014), the interviewed R&D managers indicated that open innovation involves some extra steps and additional effort in the initiation phase. This is particularly true in partner choice and the set-up of a contractual framework, both of which claim much attention and are considered the bases for smooth knowledge exchange with external partners in OI-projects (see Hoffmann and Schlosser 2001).

With respect to partner choice, it was highlighted repeatedly that the expertise and the interests (especially regarding IP rights) of all partners had to be compatible and complementary.[165] Additionally, a decent interpersonal relationship based on trust was also a strong consideration (see Hoffmann and Schlosser 2001; Whipple and Frankel 2000). Due to these important requirements for future OI-partners, companies often have to conduct a broad and extensive search to find an appropriate partner. The final partner

---

[165] To avoid conflicting interests regarding IP rights, the business models of OI-partners should not be too similar.

choice is always a strategic decision. In line with Miotti and Sachwald (2003), interviewees pointed out the purpose of an OI-project strongly influences partner choice.

Besides the right partner choice, a solid contractual framework and the associated legal security were also considered essential for knowledge exchange in OI-project. This facilitates the reduction of the organizational distance between OI-partners (see chapter 2.2.3.2) and enables the companies to claim a proportion of value generated through the OI-project (see Henkel 2006). Furthermore, the contractual framework creates security for the employees involved in the OI-project and serves as a rough guideline for the selective revealing of internal knowledge. Consequently, it helps achieve a balance between openness and secrecy or IP protection (see Hippel and Krogh 2003), which is very important for the success of an OI-project (cf. Oxley and Sampson 2004, p. 723). Only when all of the partners are willing to enter a balanced give-and-take relationship, can open innovation achieve satisfactory results for all participants. Accordingly, common ground and fairness among or between the OI-parties was considered to be another relevant aspect for knowledge exchange in OI-projects, which is in line with the findings of Hoffmann and Schlosser (2001, p. 368).

Although the contractual framework and some other formal rules serve as OI-project guidelines for the participating companies, employees mainly learn on the job. As highlighted by the interviewees, they need OI-experience to gauge appropriate ways of working with external partners and to find the right balance between give and take. Furthermore, learning on the job allows them to internalize the knowledge they are receiving from the external partner(s) and to transform explicit into tacit knowledge (cf. Cummings 2003, p. 22; Nonaka et al. 2000, p. 10). Moreover, positive experiences with an OI-partner contribute considerably to the development of mutual trust (cf. Granovetter 1985, p. 490). As already discussed in chapter 2.2.3.2 and indicated by the interviewees in the context of partner choice, trust is an indispensable component of a well-working and stable relationship (cf. Blau 1964, p. 99). It reduces the organizational and relationship distance between OI-partners and is therefore a very important enabler of knowledge exchange in OI-projects (cf. Ipe 2003, p. 347 and chapter 2.2.3.2). According to Davenport and Prusak (1998, p. 34), trust can surpass other factors' positive effect on the efficiency of knowledge exchange. Since trust develops from personal contact (cf. Davenport and Prusak 1998, pp. 35, 99), interviewees pointed to the desirability of frequent (preferably face-to-face) contact so as to reduce the physical distance between OI-partners (see chapter 2.2.3.2). In addition, many issues can be better clarified in person than by e-mail or telephone. Another basic condition mentioned by interviewees and in the literature (cf. Davenport and Prusak 1998, p. 98) is a common language, which not only refers to a mother tongue, foreign language or professional jargon. Rather, R&D managers emphasized the importance of very clear and honest communication with all OI-partners.

Besides requirements concerning the relationship and interaction of OI-partners, R&D managers emphasized several requirements on the corporate and employee levels. Consistent with Gemünden et al. (2007), they said management support and promoters that enabled employees to engage in OI-projects by granting freedom of action was vitally important. More generally, interviewees pointed to the relevance of an appropriate corporate or innovation culture that promotes and supports OI-activities. With respect to employee-related conditions for knowledge exchange in OI-projects, the R&D managers mentioned different capabilities and R&D employees' willingness to become involved. In line with the literature, the interviewees especially gave weight to absorptive capacity (see Cohen and Levinthal 1990; Zahra and George 2002). Nevertheless, communication skills, the ability to drive projects forward, open-mindedness, and sensitivity to other (corporate) cultures were also considered relevant. Relating to employees' willingness, R&D managers pointed to the necessity for the appropriate framework conditions to encourage them, i.e., the meaningfulness of an OI-project should be clearly recognizable to employees involved in the specific project. Furthermore, involvement should be associated with a personal benefit to the employees (e.g., fun, international experience).

By comparing the statements given by the R&D managers in the interviews with the employees' responses to the open-ended survey question (see Figure 26), it becomes evident the opinions regarding important conditions for knowledge exchange in OI-projects are very similar. Both groups attached the greatest importance to legal security created by a solid contractual framework (based on NDAs and a clearly defined IP rights' allocation) and to an appropriate partner choice, which allows a trusting and friendly relationship to develop; entailing open, frequent, and preferably face-to-face communication with the external partner. Associated with partner choice, R&D managers and employees also agreed on the relevance of (complementary) competencies. Nevertheless, both sides affirmed that in addition to their expertise OI-partners have to fulfill further criteria, e.g., the partner should be reliable and his/her culture and language should be compatible with the related company's corporate culture and language, where language not only refers to a common mother tongue, foreign language or professional jargon, but rather to very clear and honest communication. Other aspects considered important by both groups were fairness between the OI-parties and common ground. The necessity for a certain degree of freedom of action, where management in particular was allowed to free up space, was also mentioned by both managers and employees. However, the parties seem to attach different importance to it, with R&D employees citing it less.

Two aspects raised primarily by R&D employees related to the clarity and the added-value of OI-projects. More precisely, employees stated that project scope (including tasks, responsibilities, and milestones) and objectives have to be clearly defined and communicated. Furthermore, the project's success and the benefit of open innovation must be reasonably visible, i.e., the OI-project should appear to add value that other, "closed" innovation projects could not accomplish at that point. This statement is consistent with the belief of R&D managers in the importance of giving employees an understanding of the OI-project's rationale.

Aspects Mentioned in Interviews (by R&D Managers)	Aspects Mentioned in Survey[†] (by R&D Employees)
• Project set-up, especially partner choice (complementary expertise, personal fit, different business model) and contractual framework (NDA, IP right allocation) • Good rapport (trust, frequent and preferably face-to-face contact) • Common ground and fairness (similar attitude, cooperative behavior, balanced give-and-take relationship) • OI-experience to enable learning by doing • Common language (not only related to mother tongue, foreign language or professional jargon, but rather to very clear and honest communication) • Freedom of action through management support • Personal capabilities (absorptive capacity, communication skills, ability to drive projects forward, open-mindedness, sensitivity for other (corporate) cultures) • Personal benefits for participating in OI-project in order to facilitate employees' willingness to involve in OI-activities • Appropriate innovation culture	• Legal security (NDA , IP right allocation) • Good rapport (mutual trust, open dialogue, frequent and preferably face-to-face contact) • Common ground and fairness (common goals and interests, win-win situation) • Expertise (complementary know-how, expert in the relevant area) • Clarity (clear definition of project scope, including tasks, objectives, timeline, budget, next steps) • General partner-fit (reliable partner, no direct rival) • Added value that other or "closed" innovation projects could not accomplish • Freedom of action (availability of necessary resources, possibility to experiment) through management support [†] The aspects are ordered by their frequency of nomination (cf. Figure 24), beginning with the most mentioned aspect.

Figure 26: Relevant Aspects for Knowledge Exchange in OI-Projects[166]

The R&D managers additionally mentioned several employee-related aspects relevant to knowledge exchange in OI-projects, supporting the underlying argument of my thesis that individuals play a crucial role in the OI-context. In line with Lichtenthaler and Lichtenthaler (2009), they highlighted the relevance of absorptive capacity (see Cohen and Levinthal 1990) but also mentioned other personal capabilities such as communication skills and open-mindedness as important requirements for knowledge exchange in OI-projects. Furthermore, the R&D managers highlighted the meaning of OI-experience, which facilitates on the job

---

[166] Author's illustration

learning. In addition, positive experiences can build trust and increase employees' willingness to participate in OI-projects. Another way to encourage this willingness is the establishment of motivational framework conditions. The interviewees claimed that employees' involvement should involve personal benefits for the employees. Lastly, R&D managers aggregated some of the already discussed requirements by highlighting the need for an appropriate and supportive innovation culture.

Overall, the findings from the interviews and the open-ended survey question correspond to a large degree with the literature relating to inter-firm co-operation and customer integration (see e.g., Hoffmann and Schlosser 2001; Wecht 2006; Whipple and Frankel 2000).

7.2 RQ2: Determinants of R&D Employees' Intention to Exchange Knowledge in OI-Projects

The second research question was framed to investigate which factors determine the intention of R&D employees to exchange knowledge with external partners in OI-projects and to examine the existence of an over-riding, dominant factor. According to Ajzen's (1991) TPB, intention is determined by individuals' attitude, subjective norm, and perceived behavioral control. However, the predictive power of the factors might vary across situations and behaviors (see knowledge exchange-related studies in Table 1 and Ajzen 1985) and can rarely be forecast. Consequently, three TPB-conform hypotheses were formulated (H1, H2, H3), stating that each of the three factors (H1: attitude, H2: subjective norm, H3: perceived behavioral control) could have a positive effect on R&D employees' intention to exchange their knowledge in OI-projects (see chapter 3.3.1).

The hypotheses were tested by conducting an online survey among R&D employees and analyzing the data through variance-based structural equation modeling (PLS). The findings (see chapter 6.5) revealed that all three TPB-related hypotheses (H1, H2, H3) were indeed strongly supported by the data (see Table 19). Attitude, subjective norm, and perceived behavioral control explained 51% of intention's variance ($R^2 = 0.508$), indicating that these three factors significantly determine the intention of R&D employees to exchange knowledge with external partners in OI-projects. However, the predictive power of the determinants varied greatly. Subjective norm had by far the strongest and most significant impact on the intention of the surveyed R&D employees and so can be considered the dominant influencing factor.[167] Attitude and perceived behavioral control had a comparably strong impact on intention, even though the effect of attitude was slightly more significant. With respect to RQ2, the findings indicate that attitude, subjective norm, and perceived behavioral factors determine the intention of R&D employees to exchange knowledge with external partners in

---

[167] The link between subjective norm and intention was the strongest and most significant relationship in the entire structural model.

OI-projects. Furthermore, the results attest to the existence of a dominant factor. However, the results indicate that it is not attitude – as might be expected (see chapter 3.1.4) – but subjective norm.

In addition to inferences related to the second research question, further conclusions can be drawn from the findings from TPB components. For instance, the EFA of the perceived behavioral control construct revealed that the fragmentation of PBC into perceived self-efficacy and perceived controllability is not always necessary (in contrast, see Armitage and Conner 1999a, 1999b; Manstead and Eekelen 1998; Terry and O'Leary 1995). A second conclusion relates to R&D employees' intention to exchange their knowledge in OI-projects. In line with the literature (cf. Constant et al. 1994, pp. 404ff.), the findings from the survey show it makes a difference whether the information to be exchanged is tangible/documented or intangible/undocumented. The R&D employees indicated they were more likely to exchange undocumented knowledge with external partners than to share documented knowledge within OI-projects because of fears over confidentiality and for the sake of efficiency and information quality (see chapter 6.2). Constant et al. (1994, p. 414) further suggest that intangible knowledge reflects an employee's identity, qualities, and value. Therefore, the exchange of intangible knowledge is a way of imparting their expertise and might contribute to the feeling of pride and belonging. A third conclusion, which was derived from the indicator loadings and indicator weights of the formatively measured subjective norm construct (see Table 17), concerns the absolute and relative importance of different groups of people for R&D employees' intention to exchange knowledge with external partners (cf. Hair et al. 2011, pp. 145f.). The findings showed that social pressure caused by the CEO (SN1) and colleagues (SN3) both had high absolute and relative importance. In contrast, the subjective norm related to the immediate supervisor (SN 2) only had a significant absolute importance, since the level of perceived social pressure caused by the immediate supervisor was generally high. Consequently, the marginal effect of social pressure caused by the immediate supervisor is lower than the marginal effect of social pressure caused by the CEO or colleagues. My last conclusions result from the link between the TPB components and the individual-related OI-barriers identified in chapter 2.1.3.5. As already explained in chapter 3.1, the "want-barrier" is often due to a negative attitude, the "shall-barrier" can be considered the result of subjective norm and the "can-barrier" might be associated with a lack of perceived behavioral control (cf. Behrends 2001, p. 96; Enkel 2009, pp. 189ff.; Haller 2003, pp. 192ff.; Hauschildt and Salomo 2011, pp. 125f.). Accordingly, the dominant influence of subjective norm suggests the "shall-barrier" might entail the biggest obstacle to knowledge exchange in OI-projects, followed a considerable distance behind by the "want-barrier" and "can-barrier". This implies measures to reduce the "shall-barrier" will likely show the strongest positive impact on employees' knowledge exchange behavior in OI-projects, while efforts to reduce the "can-barrier" – such as training (cf. Cabrera and Cabrera 2002, pp. 700f.) – might have the poorest effect. Furthermore, it weakens the assumption that the NIH-syndrome, which reflects negative attitude toward external knowledge (cf. Clagett 1967, p. ii) and is considered as an important manifestation of the "want-barrier", is a major obstacle to open innovation (see chapter 2.1.3.5). According to my findings, the "want-barrier" and the NIH-syndrome as they affect individuals may not be the most substantial problems for OI-attempts in participating companies. However, the strong impact of subjective norm on R&D employees' intention implies the NIH-syndrome could possibly become a great problem as it not only affects individuals but groups of people. If a critical mass of people or a group of relevant individuals have negative attitude toward knowledge exchange in OI-project they might form a subjective norm. Furthermore, a snowball effect could extend this collective NIH-syndrome by "infecting" other individuals with the NIH-syndrome. However, even if the NIH-syndrome theoretically poses a high risk to OI-projects, neither the interviews nor the survey results highlighted its presence in any acute way in the R&D departments of the participating companies. The interviewed R&D managers said their employees had predominantly positive opinions on open innovation. R&D employees would only have a clear preference for closed innovation if they had encountered negative experiences with open innovation or if the results of a "closed" innovation process were as promising as the outcome of an OI-process. This statement was strongly supported by the descriptive results relating to employees' attitude. As shown in Figure 18, the surveyed R&D employees have a (very) favorable attitude toward their knowledge exchange with external partners in OI-project.[168]

7.3 RQ3: Motivational Factors with Positive Influence on R&D Employees' Willingness to Exchange Knowledge in OI-Projects

The third research question was framed to investigate which motivational factors positively influence R&D employees' willingness to exchange their knowledge with external partners in OI-projects. Since willingness is closely related to the TPB's attitude construct, I conducted a review of TPB-related studies to identify motivational factors that would presumably influence individuals' attitudes toward exchanging their knowledge in OI-projects. Due to the absence of literature combining the TPB (or at least the TRA) with research on open innovation or knowledge exchange in OI-projects, I used articles that investigated individuals' knowledge exchange by means of the TPB/TRA as a proxy. In the course of the literature review, four motivational factors frequently emerged: enjoyment in helping, sense of self-worth, reciprocity, and rewards. However, the pre-test of the questionnaire suggested a differentiation between two types of rewards. Consequently, five hypotheses relating to

---

[168] A possible explanation for the absence of the NIH-syndrome could be the employees' lengthy and predominantly positive experiences with open innovation, which might have cured the NIH-syndrome in the R&D departments of the participating companies.

motivational factors and their influence on R&D employees' attitude toward knowledge exchange in OI-projects were finally formulated (H4, H5, H6, H7a, H7b; see chapter 3.3 and 4.4.2). Four of the five hypotheses assumed a positive relationship between the respective motivational factor and attitude (H4: enjoyment in helping, H5: sense of self-worth, H6: reciprocity, H7b: reward B). On the contrary, reward A was expected to negatively influence attitude (H7a).

As in the case of the TPB-related assumptions, the hypotheses referring to the motivational factors were tested by conducting an online survey among R&D employees and analyzing the data through variance-based structural equation modeling (PLS). The findings (see chapter 6.5) revealed that three of the five hypotheses were supported by the data (see Table 19). As expected, enjoyment in helping (H4), sense of self-worth (H5), and reward B (H7b) had a significant and positive impact on R&D employees' attitude. They explained 31% of attitude's variance (R^2 = 0.313), indicating that these three motivational factors considerably determine the attitude of R&D employees toward knowledge exchange with external partners in OI-projects. However, the predictive power of the determinants varied. Enjoyment in helping had by far the strongest and most significant impact on R&D employees' attitude. Thus, it can be considered the dominant motivational factor in the given context. Reward B and sense of self-worth had a comparatively strong and significant impact on attitude.

Contrary to my expectations, neither reciprocity (H6) nor reward A (H7a) significantly influenced R&D employees' attitude to exchanging knowledge with external partners in OI-projects; reciprocity was not positively related to attitude nor did reward A have a negative effect on attitude. Although unexpected, the lack of association between reward A and attitude did not come as a complete surprise (see chapter 3.3.5). The literature review had already indicated a very ambiguous picture (see Table 2). In contrast to reward B, reward A represents a hygiene factor and, consequently, measures dissatisfaction or no dissatisfaction rather than no satisfaction or satisfaction, i.e., the absence of reward A (negative) leads to dissatisfaction, while its presence (positive) results in no dissatisfaction (see Figure 27 and Herzberg et al. 1959; Herzberg 1968, 1974).[169]

---

[169] Accordingly, the absence of reward B (negative) leads to no satisfaction, while the presence (positive) results in satisfaction.

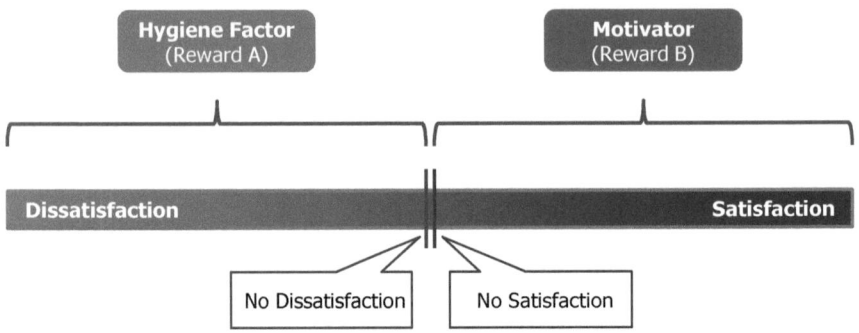

Figure 27: Herzberg's Motivation-Hygiene Theory and Reward Constructs[170]

Motivators – as the name suggests – have the potential to motivate employees by their presence. This could be shown in the study: reward B positively influenced R&D employees' attitude. The presence of hygiene factors, on the other hand, only has the potential to avoid employee demotivation. This fact offers an explanatory approach for the non-existing negative effects from reward A in this study. The absence of reward A (hygiene factor) would lead to dissatisfaction and, thus, very likely have a negative effect on employees' attitude. However, the presence of reward A would lead to no dissatisfaction, which is a rather neutral position. In this situation, it should be expected that reward A would not have an influence on employees' attitude. In this study, reward A was operationalized using items covering work assignments, promotion, salary, bonus, and reputation (see chapter 6.4.1.1 (g) in combination with Table 5). The interviews conducted with the R&D managers suggest such working conditions are at a very high level within the respective companies and so can be considered to be present. The interviewees also stated that even though involvement in OI-activities might be part of the objective agreement for an employee and, consequently, might be regarded as being relevant for the calculation of bonuses, it would not necessarily lead to a disastrous drop in payment if the employee did not optimally engage in OI-projects because the final bonus figure is affected by the realization of multiple targets. Consequently, the non-existence of a relationship between reward A and employees' attitude might be attributable to the high-level presence of this hygiene factor and so to reward A's weak demotivational impact.

Regarding the unexpected result of reciprocity being not positively related to R&D employees' attitude, the follow-up group discussions and a closer look at the literature offered several possible explanations. Some participants in the follow-up discussion at Company A contributed statements that also relate to Herzberg's motivation-hygiene theory (see Herzberg et al. 1959; Herzberg 1968, 1974). They suggested that reciprocity in the

---

[170] Author's illustration (a simplified view of Herzberg's motivation-hygiene factor theory and its connection to the reward constructs of this study)

context of inter-organizational knowledge exchange represents also a hygiene factor rather than a motivator. R&D employees take a balanced give-and-take relationship for granted, particularly because reciprocity is institutionalized and regulated through the contractual framework of the OI-project. Furthermore, the R&D employees rely on management's ability to select OI-partners on the basis of their willingness to enter a balanced give-and-take relationship. Consequently, the absence of reciprocity may cause dissatisfaction but the presence of it does not satisfy or motivate the R&D employees. One of the participants in the follow-up discussions compared knowledge exchange in OI projects with money exchange in a currency exchange office: When a person enters the currency exchange office to exchange euros to dollars, he/she expects to receive dollars for euros, i.e., it is not a motivator but rather a basic requirement or hygiene factor to receive dollars for euros. If the person does not receive dollars, he/she will be dissatisfied with the exchange. However, if the person receives the dollars as expected, he/she will not be satisfied but only not dissatisfied – as this was no more or less than the person expected. The same applies to knowledge exchange in OI-projects: If an R&D employee joins an OI-project and exchanges knowledge, he/she will provide knowledge not because he/she is motivated by reciprocity, but because he/she subconsciously considers it a basic requirement. Consequently, knowledge return will not motivate or positively influence his/her attitude toward knowledge exchange in OI-projects. However, the absence of a return will lead to dissatisfaction and could have a negative influence on employees' attitude. A second explanatory approach derives from the literature and is related to unequal positions of power among the OI-partners. Blau (1964, pp. 104f.) and Gouldner (1960, pp. 164ff.) pointed out that an imbalance in power can interfere with reciprocity, since the stronger partner might not always provide a fair return and so exploit the weaker party. This effect is intensified if the weaker partner does not have a suitable alternative and cannot simply choose another OI-partner. The interviews showed that even large and multinational companies might find themselves as the weaker partner. The R&D managers repeatedly stated their eagerness to co-operate with a luminary in a specific field. Consequently, they might not always be very flexible with respect to partner choice, which can create a dominant position of power on the part of the favored partner. In such a case, employees would possibly exchange their knowledge even if a reciprocal relationship wasn't expected. However, an arrangement based on this kind of imbalance is less stable than a relationship entailing reciprocity. This is in line with the argument from the follow-up discussion that most partnerships are durable and that reciprocity plays a crucial role. A third direction that could help to explain why reciprocity did not show any effect on R&D employees' attitude is suggested by Wasko and Faraj (2000). They found indications for the existence of a generalized form of reciprocity in the sense that a person providing knowledge to somebody else does not expect a return from a direct counterpart. The reciprocity construct in this study was operationalized accordingly (see Table 5). However, the generalization of reciprocity might not only apply to the individuals involved in a reciprocal relationship but could possibly also be extended to the purpose of exchange, i.e., R&D employees providing their knowledge might not necessarily expect to get back knowledge but something else (e.g., praise, recognition, status). This aspect is not covered by the reciprocity construct used in this study. Consequently, this form of reciprocity is not measured: Its influence on attitude is not evaluated in my study. A last explanatory approach refers to the interdependence of individuals' behavior. Following Kelley and Thibaut (1978, pp. 282ff.), the behavior of individuals acting independently from one another differs from the behavior of individuals influenced by their social and organizational context. With respect to knowledge exchange, Constant et al. (1994, pp. 401f.) interpreted the interdependence theory of Kelley and Thibaut – which is closely related to social exchange theory – by reasoning that if individuals acted independently from their counterparts, self-interest and reciprocity would be the most important predictors of behavior. However, if individuals acted in a social and organizational context that supported pro-social behavior, people might not have negative reciprocity, i.e., even though person A does not share his/her knowledge with person B, person B would share his/her knowledge with person A. This argument further strengthens the finding that subjective norm has a strong impact on R&D employees' intention to exchange knowledge.

With respect to RQ3, the findings indicate that enjoyment in helping, sense of self-worth, and reward B can positively influence R&D employees' willingness to exchange their knowledge with external partners. However, reciprocity and reward A do not have a positive impact on R&D employees' attitude toward knowledge exchange in OI-projects. This outcome supports the differentiation between job-context related, rather extrinsic factors (hygiene factors) and job-content related, rather intrinsic factors (motivators) suggested by Herzberg (1968; 1974). Reward A and reciprocity, which did not show a positive influence on R&D employees' attitude, could be classified as hygiene factors. Enjoyment in helping and in particular sense of self-worth and reward B are typical motivators according to the motivator-hygiene theory.[171] The interviews support this conclusion by underlining the relevance of intrinsic motivation in OI-projects. The R&D managers stated their employees showed a high intrinsic motivation to solve tricky and challenging tasks, which they put down to their eagerness to experiment. Since open innovation is often applied in cases where the task is too complex and too difficult for a single player, these projects offer interesting tasks and a certain degree of intellectual freedom, both of which are great stimuli for R&D employees. Consequently, they focus on solving the problem rather than on its contextual aspects (i.e., they don't care if they solve the task with or without an external partner).

---

[171] The personal increase of knowledge is a notable motivational factor among those surveyed. This becomes evident by looking at the descriptive results of attitude and rewards (see Figure 18 and Figure 20). The highest-rated attitude-item is: "My knowledge exchange with external partners in OI-projects is (very) valuable to me" and the highest-rated reward item is: "When I exchange knowledge with external partners in OI-projects it is important for me to increase my knowledge".

8 Conclusions

This chapter considers the findings of my study with regard to their contribution to academic research. Furthermore, managerial implications are derived and recommendations for managerial practice are formulated. Although I executed my research with great care and thoroughness, it is inevitably subject to some limitations, which are highlighted in the last sub-chapter together with recommendations for further research.

8.1 Contribution to Academic Research

First and foremost, my thesis significantly contributes to OI-research. It is the first study applying the TPB in the context of open innovation and relating it to barriers at the individual level. Moreover, the study links open innovation to other research fields such as knowledge management and motivation theory. In so doing, my findings also make a contribution to knowledge exchange and motivation research. Overall, this study broadens the view on open innovation and substantially contributes to the current OI-understanding.

8.1.1 Contribution to Open Innovation Research and the TBP

My thesis contributes to OI-research in a variety of ways. Previous studies have emphasized the organizational level and rarely considered the people side of open innovation. Those studies that did concentrate on the individual level mainly examined lead users and individuals engaged in open-source projects or other OI-communities. My thesis is the first empirical and quantitative study with a clear focus on R&D employees who work for an OI-embracing company and have experience with OI-projects. By challenging the dominant position of the organizational level in OI-studies and targeting a set of relevant questions related to the human side of open innovation, my thesis significantly contributes to the micro-foundation of OI-research. It examines open innovation from the seldom adopted R&D point of view and sheds light on the hitherto neglected perspective of individuals engaged in OI-projects.

It is also the first time Ajzen's TPB has been applied in the context of open innovation and that determinants of R&D employees' intention to exchange knowledge with external partners in OI-projects have been the focus of analysis. Furthermore, my study is the first to link the components of the TPB to the three individual-related barriers suggested by Behrends (2001, p. 96) and Hauschildt and Salomo (2011, pp. 125f.). This combination allows meaningful conclusions on the relevance of different barriers. The results of the online survey indicated that perceived social pressure (subjective norm) and therefore the "shall-barrier" had an immense impact on R&D employees working in OI-projects. Indeed, it had the strongest impact by far. The formative measurement of subjective norm allowed the investigation of the absolute and relative importance of the three categories (CEO, immediate supervisor, and colleagues) that were expected to represent all possible sources of social pressure in a professional context (cf. Karahanna et al. 1999, p. 201). The data revealed that the tested groups differed in their absolute and relative importance and, thus, contributed differently strong to the social pressure. The CEO and his/her beliefs and the peer group of R&D employees represented by their colleagues had high absolute and relative importance, while the immediate supervisor possessed only an absolute importance. With respect to attitude, the findings implied that R&D employees' attitude toward knowledge exchange in OI-projects – regardless of whether these are positive or negative – do not play the dominant role in predicting the intention to exchange knowledge with external partners in OI-projects. This undermines the assumption that "want"-related barriers such as the NIH-syndrome play the dominant role in knowledge exchange in OI-projects. The results related to the perceived behavioral control indicate its effect on intention is comparable to attitude's impact on intention. Furthermore, the data showed that the splitting of the construct into self-efficacy and behavioral control claimed by several scholars (see Armitage and Conner 1999a, 1999b; Manstead and Eekelen 1998; Terry and O'Leary 1995) is not always necessary. Even though the construct of perceived behavioral control definitively requires both aspects, my findings did not confirm the requirement to treat them as two independent variables.

The interviews and the answers to the open-ended survey question provide valuable insights to the mindset of R&D employees and managers. In particular, the findings uncovered that from an R&D perspective the most important requirements for participating in knowledge exchange in OI-projects relate to legal security, the selection of an external partner, and the building of a trusting relationship with this party, entailing common ground and fairness. These findings are in line with studies examining the success factors of strategic alliances (see Hoffmann and Schlosser 2001; Whipple and Frankel 2000). This confirms that OI-research relates to research on inter-organizational co-operation and suggests results may also be comparable and transferable to a certain extent. My thesis not only substantiates this link; it also relates OI-research to knowledge management and motivation theory. It provides an indication of motivational factors derived from the knowledge exchange literature that has relevance to open innovation. The contribution of my study to these two research fields are discussed in the following

> *"We also support initiatives that aim to couple open innovation to other disciplines or management areas." (Vrande et al. 2010, p. 231)*

8.1.2 Contribution to Knowledge Management Research

This study contributes to knowledge management research by compiling a comprehensive knowledge management process based on the existing literature (see Figure 7). This identifies the most OI-relevant phase of this process and relates knowledge management to open innovation. The study challenges the applied terminologies for this OI-relevant phase (i.e., knowledge exchange, knowledge sharing, and knowledge transfer) and adopts the term "knowledge exchange" after careful deliberation (see chapter 2.2.3).

Due to the lack of appropriate OI-literature, studies investigating knowledge exchange in different contexts by means of the TPB were used as a proxy to derive motivational factors with a positive influence on R&D employees' attitude. The vast majority of these studies were conducted in Asian countries. Consequently, my study contributes to knowledge management research by adding an analysis conducted primarily in Europe. The findings of my thesis show that some of the motivational factors derived from the knowledge exchange literature have a significant impact on employees' attitude toward knowledge exchange in OI-projects, confirming the connection between open innovation and knowledge exchange.

Last but not least, the study uncovers the importance of differentiating between the exchange of documented and undocumented knowledge in the context of open innovation. The data clearly shows the intention to exchange undocumented knowledge is much more pronounced than the intention to exchange documented knowledge. The follow-up group discussions provide an explanatory approach for this finding by highlighting the differences of documented and undocumented knowledge in terms of confidence, efficiency, and information quality.

8.1.3 Contribution to Motivation Theory

The findings of my study strongly support Herzberg's (1968; 1974) motivation-hygiene theory, i.e., the difference between the job-content related, rather intrinsic factors that lead to job satisfaction (motivators) and job-context related, rather extrinsic factors that lead to job dissatisfaction (hygiene factors). Since this distinction is rarely considered in the context of knowledge exchange and/or OI-literature, my thesis makes a contribution by broadening the scope of motivation theory.

In addition, the discovery that the surveyed R&D employees are mainly intrinsically motivated to exchange their knowledge with external partners in OI-projects and collaboratively solve the given problems should be noted. All tested motivational factors that were found to significantly and positively influence R&D employees' attitude toward exchanging their knowledge in OI-projects (i.e., enjoyment in helping, sense of self-worth, and reward B) address the intrinsic interests of employees. Since companies draw mainly on the OI-approach to solve problems that are very complex, difficult, and/or novel, OI-projects tend to offer challenging tasks and a certain degree of freedom. The conclusion that R&D employees engaged in OI-projects are mainly intrinsically motivated appears intuitive and in line with the arguments of Ryan and Deci (2000, pp. 59f.) *"[...] that intrinsic motivation will occur only for activities that hold intrinsic interest for an individual—those that have the appeal of novelty, challenge, or aesthetic value for that individual."*

In contrast to the three intrinsically oriented motivational factors, reward A and reciprocity did not show any relevant effect on R&D employees' attitude. These two factors were classified as hygiene factors in the context of this study for different reasons (see chapter 7.3). The finding that (the rather extrinsic) reward A did not had an impact on attitude, while (the rather intrinsic) reward B positively influenced attitude showed intrinsic reward top extrinsic reward in the context of open innovation in R&D departments. Following Jewels and Ford (2006, p. 108) *"[...] knowledge workers are less likely than traditional workers to be motivated by extrinsic rewards."* My thesis therefore contributes to motivational theory by confirming the importance of distinguishing among different kinds of rewards (i.e., intrinsic and extrinsic) – particularly in the context of knowledge exchange in OI-projects. The rewards construct(s) should therefore be operationalized accordingly. Moreover, my study further contributes by introducing a new reward construct (reward B), which entails intrinsic elements. This was established through liaison with R&D employees during the pre-test and successfully applied in this study.

8.2 Managerial Implications

The findings of my study show that attitude ("want"), subjective norm ("shall") as well as perceived behavioral control ("can") influence R&D employees' intention to exchange their knowledge in OI-projects significantly – though the perceived social pressure had by far the strongest influence. In addition, the study reveals the dominant role of intrinsic motivation and indicates the distinction of motivators and hygiene factors is very important in understanding employee motivations. Furthermore, the interviews and answers to an open-ended survey question shed light on the basic requirements for knowledge exchange between R&D employees and external partners in OI-projects. From all these findings I can draw conclusions relevant to academic research. Moreover, my thesis also entails several implications for managers of companies already engaged in OI-activities and for OI-newcomers. Figure 28 provides an overview of recommendations for managerial practice based on the findings of my thesis.

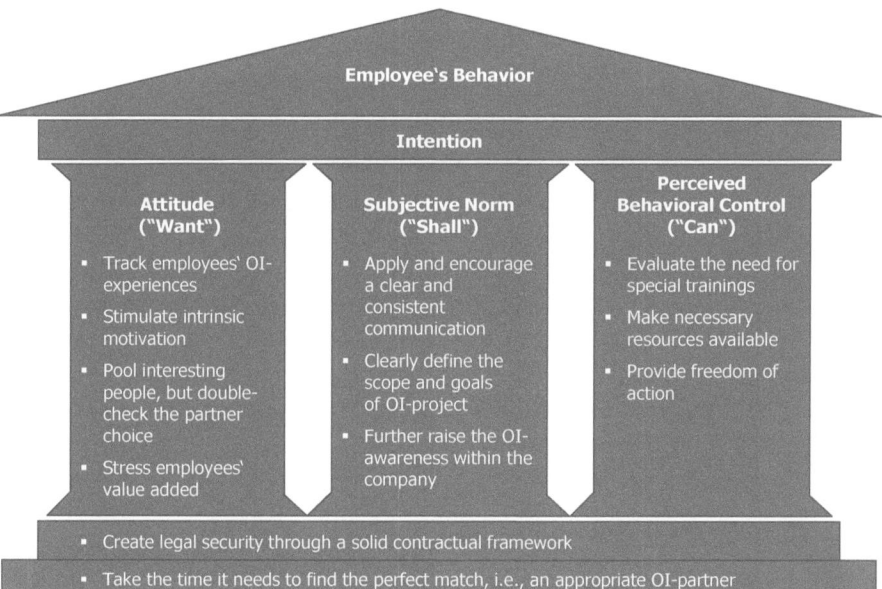

Figure 28: Recommendations for Managerial Practice along TPB Components[172]

As indicated by the interviews, the correct choice of partner is at the foundation of any promising OI-project. An appropriate OI-partner should possess complementary and compatible expertise. Furthermore, a good rapport and trusting relationship based on common ground and fairness should exist beforehand or have the potential to develop over the period of project co-operation. Consequently, managers should take enough time to find a suitable partner because the professional and personal fit of the partner is a basic requirement for a successful OI-project. The time managers think they save through a fast initial selection can easily cost them time in the end if the partner turns out to be the wrong choice. Furthermore, the selection must be based on rational criteria, e.g., the selection should not be made out of courtesy or because it seems like an easy choice.

After the perfect match is found, great value has to be attached to the creation of legal security for all the parties (cf. Slowinski and Sagal 2010, pp. 43f.) by way of a solid contractual framework. As a basic requirement each party has to sign an NDA and the IP rights allocation has to be stipulated. This helps the partners to protect their knowledge and to claim a proportion of the value generated through the OI-project. The contractual framework should also detail the expected contributions from the parties in a reciprocal relationship. It creates security for the employees involved in the OI-project and serves as a

---

[172] Author's illustration (with reference to Figure 11)

guideline. However, the contractual framework can only act as a guideline if the basic facts are communicated to the R&D employees engaged in the OI-project. An internal meeting ahead of the official OI-project start can help to clarify the key points of the upcoming co-operation. However, this meeting should not replace the inaugural meeting involving all the participating partners (cf. Slowinski and Sagal 2010, pp. 44f.).

After the foundation of an OI-project is built and the conditions for R&D employees' behavior are set (see Figure 5), managers should take appropriate measures to positively influence the R&D employees' contribution to the OI-project. The findings of my study demonstrate it is worthwhile to consider all three aspects (i.e., attitude, subjective norm, and perceived behavioral control) that have a positive impact on R&D employees' intention to exchange their knowledge with external partners in OI-projects. However, special attention should be paid to the implications related to subjective norm because it was found to have by far the strongest influence on the intention of the surveyed R&D employees. Therefore, it has the greatest leverage effect and potential.

8.2.1 Recommendations Related to Attitude ("Want")

Attitude tends to develop from past experiences (cf. Fishbein and Ajzen 1975, pp. 9f.). In order to maintain a positive attitude or convert a negative attitude into a positive or at least neutral one, it is critical to know which aspects or conditions might have turned employees' past engagements in OI-projects into a positive or negative experience. Consequently, it is advisable to track employees' OI-experience in order to identify disruptive factors. One way to find out about these factors would be to (anonymously) survey employees or to arrange "lessons learned" sessions after every OI-project. Such methods could reveal determining factors and conditions that need to be reviewed and possibly adapted.

A second recommendation refers to the findings related to the R&D employees' motivations. Intrinsic incentives (e.g., personal development) were found to have a much stronger impact on R&D employees than extrinsic, often monetary incentives (e.g., higher salary, bonus, job security). Therefore, it is advisable to establish conditions that stimulate intrinsic motivation. Furthermore, the pre-test results suggest employees are very capable of identifying motivational tools. The incentives mentioned by the four R&D representatives at the participating companies during the pre-test very well reflected the opinions of R&D employees who participated in later surveys. The construct resulting from the pre-test (reward B) showed a significant impact on attitude. Consequently, managers should listen carefully to their employees if they want to recognize the factors that have the most potential to motivate and which are only hygiene factors.

A third recommendation refers to the already-mentioned foundation for a promising OI-project: partner choice. Broadening the own horizon (reward B) and the enjoyment in helping others are two factors that were found to be positively related to R&D employees' attitude toward exchanging their knowledge with external partners in OI-projects. The management could assist here by pooling interesting people who can learn from each other and inspire one another. However, a stimulating relationship is only possible where the partners are matched professionally and personally. It is therefore worth double-checking partner choice; a poor fit can damage employees' motivation.

A last piece of attitude-related advice refers to R&D employees' contribution to the project or company success. My findings showed that R&D employees are not only interested in helping other people; they also want to provide value to their company (sense of self-worth). Consequently, it is essential to set their engagement in a broader context and to concretely highlight where they add value to the project or the company through their knowledge exchange in OI-projects. It is advisable to point out the benefits of open innovation for the individual employee and for the company. If R&D employees recognize the relevance of the OI-project to the company's success, they are likely to perceive their task and related efforts as meaningful. This will have a positive effect on their sense of self-worth, which in turn positively influences their attitude toward their knowledge exchange in the OI-project.

8.2.2 Recommendations Related to Subjective Norm ("Shall")

As mentioned at the beginning, the perceived social pressure caused by the CEO, immediate supervisor, and colleagues had by far the strongest influence on R&D employees' intention and should therefore be explicitly taken into account.

Subjective norm consists of two components (cf. Ajzen 1985, p. 14, 1991, p. 195): On the one hand, perceived opinions and interests from a CEO, immediate supervisor, and colleagues play an important role. Thereby, employees' perception does not necessarily reflect the actual opinion of the referents (cf. Ajzen and Fishbein 1980, p. 57). On the other hand, employees' motivation to comply with the interests and wishes of a CEO, immediate supervisor, and colleagues is a crucial factor. However, employees can only act in a certain way if the true opinions of these groups are known. To minimize the gap between perceived and existing interests and so to avoid a "misdirection" of subjective norm, clear and consistent communication is required. This is not only true for messages from the CEO and other supervisors; the communication within teams should also be clear and consistent to ensure the real interests of colleagues are recognized. If employees' perception and reality diverge, it is often because the employees do not receive sufficient feedback (cf. Gecas 1982, p. 6). Consequently, it is crucial to give employees frequent feedback – both positive and negative. Positive feedback will encourage employees' knowledge exchange, while negative feedback can help to control the quality of employees' contributions (cf. Cabrera and Cabrera 2002, p. 699). Furthermore, the demand for clear and consistent communications is closely connected to the requirement mentioned by the R&D employees to clearly define the scope and goals of an OI-project.

A second approach to taking advantage of the strong positive impact of subjective norm on R&D employees' intention to exchange knowledge in OI-projects is to further raise the OI-awareness within the company so as to establish OI-promoters (see Gemünden et al. 2007) on every hierarchical level and to increase the group of OI-interested people. Therefore, the added value of OI-projects should be emphasized without neglecting possible obstacles. Reports about experiences, concrete achievements, and lessons learned (e.g., using the intranet or in roadshows) could serve as a good starting point.

8.2.3 Recommendations Related to Perceived Behavioral Control ("Can")

Even though the perceived behavioral control showed the weakest positive effect on R&D employees' intention to exchange knowledge in OI-projects, related measures have the potential to positively influence their intention. Consequently, it is also essential that management supports its R&D employees by giving them space and helping them feel capable of coping with tasks and challenges related to an OI-project.

A starting point should be necessary training. Managers should evaluate the need for special training on a regular basis. Communications and legal training, for instance, might be required in the context of open innovation. Furthermore, knowledge-enhancing training and employees' participation at conferences can help improve absorptive capacity, which is also desirable in an OI-context.

If management aims to increase R&D employees' perceived control over their knowledge exchange in OI-projects, existing freedom of action could be reviewed. In particular, it should be ascertained whether employees have all resources at their disposal that are relevant for an effective knowledge exchange (e.g., a budget for training and business trips to meet external partners, enough time for training and to engage in OI-projects).

The implications and recommendations for managerial practice are addressed to managers of both OI-active companies and OI-newcomers. They can learn how to leverage R&D employees' intention to exchange their knowledge in OI-projects. Furthermore, both can use the results to reconsider their incentive systems. The findings of my study can also provide OI-newcomers with an overview of the most important requirements for knowledge exchange in OI-projects (e.g., a proper selection of OI-partners and legal security). Managers of companies already following the OI-approach can use the results to reflect if, and to what extent, they comply with the requirements.

8.3 Limitations and Suggestions for Further Research

Although I was diligent with my research design and executed the research carefully, the study is subject to some limitations that need to be recognized and considered when interpreting my results. The limitations mainly arise from characteristics of the sample and from the research design. I will outline them in the following and make suggestions for related further research. Subsequently, I will highlight some additional starting points for further research that became evident to me in the course of my research.

The first limitation results from my thesis' contribution to academic research. As it was the first study to focus on open innovation in R&D departments and on R&D employees exchanging their knowledge with external partners in OI-projects, further comparable analyses need to follow to confirm my findings. The survey sample was compiled using R&D employees from four manufacturers. These are all global businesses headquartered in Germany; active in the B2B market; operate in the fields of chemistry, automation, and steel treatment; and publicly state their support for the OI-approach. Even though this given mix of characteristics might be representative of many (high-tech) industries and companies, my findings should be interpreted in the described context and other characteristics might mean different results. The interviews I conducted with R&D managers suggest OI-projects and OI-culture differ in B2B and B2C environments and across continents (the interviewees highlighted in particular the apparent differences in open-mindedness found in America and Europe). Consequently, further studies in different contexts (e.g., the fast-moving consumer goods industry, the B2C market, American companies) are required to analyze which findings regarding the impact of attitude, subjective norm, and perceived behavioral control on R&D employees' intention to exchange knowledge in OI-projects are general and which are specific.

A second limitation might be the sample size. The number of usable responses was adequate for testing the research model and related hypotheses but it could be dangerous to extrapolate my findings. Also, the sample size was not sufficient to independently calculate the research model for different groups and to conduct valid group comparisons.[173] Even though such analyses had not been the focus of my research, they could have provided interesting insights. The investigation of differences between different companies, industries, levels of OI-experience, and the "Big Five" main personality traits in particular might have been worthwhile. For that reason, future studies might aim for a higher sample size to allow examination of the disparity of groups.

---

[173] As already mentioned in chapter 4.4.5, the rule of thumb suggested by Chin 1998b, p. 311 implies that my research model requires a minimum of 50 observations. If the sample size is below this minimum, PLS cannot provide robust outcomes.

The final limitations result from the research design. Due to reasons of feasibility (see chapter 3.1.5), the behavior construct of TPB was not part of my research model. Future research might investigate the relationship between intention and behavior and explore the stability of this connection in the context of knowledge exchange in OI-projects. So as not to reduce my study to a specific case of open innovation, the online survey did not question R&D employees with regard to particular OI-projects or OI-partners. Nevertheless, the employees were asked to state the frequency of co-operation with different partners (see Figure 16). The answers to this question plus the R&D employees' comments[174] imply the types of OI-projects they were usually engaged in. However, a definite conclusion cannot be drawn. Consequently, future research could conduct a case-by-case analysis to evaluate the correlation between R&D employees' opinions and specific OI-project characteristics. A last limitation originates from the application of self-report measures. However, since my research focused on individual R&D employees, it was only reasonable to directly consult this target group. Thus, I carefully applied measures for bias treatment (see chapter 6.1.2) to minimize the possible effect.

Generally, OI-research should increasingly focus on the individuals engaged in OI-projects to broaden the understanding of the most fundamental level of open innovation. My study was only able to cover some factors influencing R&D employees engaged in OI-projects. However, there are plenty of aspects that might have a great impact: Corporate or innovation culture, for instance, was mentioned by the interviewed R&D managers and was also considered relevant by other researchers (see Herzog 2008). Furthermore, governance is a crucial topic in the context of open innovation and knowledge exchange; this also deserves more attention in future empirical research (see Foss et al. 2010; Grandori 2001). Another aspect with great relevance for knowledge exchange in OI-projects is dual allegiance (see Gordon and Ladd 1990). According to Husted and Michailova, R&D employees engaged in OI-projects experience pressure to be loyal and to have an obligation to both their company and the OI-project:

> *"In the context of dual allegiance R&D workers need to constantly decide what knowledge to share and when, with whom, and to what extent, in order to be loyal to the organization that employs and pays them. At the same time, they are also allegiant to the collaboration, as they need to play a meaningful role in it and add value to it." (Husted and Michailova 2010, p. 38)*

Even though this issue has already found its way into OI-research and had been discussed in connection with inter-organizational R&D collaboration and open source software companies (see Chan and Husted 2010; Husted and Michailova 2010), it might be worth expanding the

---

[174] At the end of the online survey employees were asked to leave some comments or feedback. Several employees used this comment box to clarify their answers and some of them stated their usual OI-partner.

OI-research in this area. A possible starting point might be to examine if and how governance mechanisms can support employees by dealing with dual allegiance.

Since my study combined open innovation and knowledge exchange, I can also derive recommendations for further research concerning the knowledge component in open innovation. One suggestion would be to relate OI-research to other knowledge management processes and elements. My thesis only considered the exchange of knowledge, i.e., its give and take. It did not assess whether employees were able to absorb knowledge from external sources. Consequently, future studies could investigate the absorption of external knowledge and its integration into the internal innovation process. In addition, future research could draw a distinction between formal and informal knowledge exchange and examine the role of both in the context of open innovation (cf. Alavi and Leidner 2001, pp. 120f.).

A last starting point for further research that struck me during the interviews with R&D managers refers to the measurability of open innovation and its success and productivity in particular. Even though interviewees could tell which OI-projects were relatively successful and which were not, their answers implied a lack of measurable key figures and key performance indicators relating to open innovation. This is in line with Chesbrough (2006c, p. 10; 2006a, p. 20) who said former R&D metrics (e.g., number of patents and publications generated, percentage of sales invested in internal R&D) are outdated in the context of open innovation and should be revised.

—

References

Afuah, Allan; Tucci, Christopher L. (2012): Crowdsourcing as a Solution to Distant Search. *Academy of Management Review* 37 (3), pp. 355–375.

Ajzen, Icek (1985): From Intentions to Actions: A Theory of Planned Behavior. In Julius Kuhl, Jürgen Beckmann (Eds.): Action Control. From Cognition to Behavior. Berlin: Springer (Springer Series in Social Psychology), pp. 11–39.

Ajzen, Icek (1991): The Theory of Planned Behavior. *Organizational Behavior & Human Decision Processes* 50 (2), pp. 179–211.

Ajzen, Icek (2002a): Constructing a TpB Questionnaire: Conceptual and Methodological Considerations.

Ajzen, Icek (2002b): Perceived Behavioral Control, Self-Efficacy, Locus of Control, and the Theory of Planned Behavior. *Journal of Applied Social Psychology* 32 (4), pp. 665–683.

Ajzen, Icek; Fishbein, Martin (1980): Understanding Attitudes and Predicting Social Behavior. Englewood Cliffs, NJ: Prentice-Hall.

Ajzen, Icek; Fishbein, Martin (2005): The Influence of Attitudes on Behavior. In Dolores Albarracin, Blair T. Johnson, Mark P. Zanna (Eds.): The Handbook of Attitudes. Mahwah, NJ: Lawrence Erlbaum Associates, pp. 173–221.

Ajzen, Icek; Madden, Thomas J. (1986): Prediction of Goal-Directed Behavior: Attitudes, Intentions, and Perceived Behavioral Control. *Journal of Experimental Social Psychology* 22 (5), pp. 453–474.

Alavi, Maryam; Leidner, Dorothy E. (2001): Review: Knowledge Management and Knowledge Management Systems: Conceptual Foundations and Research Issues. *MIS Quarterly* 25 (1), pp. 107–136.

Albers, Sönke; Brockhoff, Klaus K.; Hauschildt, Jürgen (Eds.) (2000): Betriebswirtschaftslehre für Technologie und Innovation. Graduiertenkolleg; Eine Leistungsbilanz. Kiel: Institut für Betriebswirtschaftliche Innovationsforschung der Christian-Albrechts-Universität zu Kiel.

Allen, Robert C. (1983): Collective Invention. *Journal of Economic Behavior & Organization* 4 (1), pp. 1–24.

Allen, Thomas J. (1977): Managing the Flow of Technology. Technology Transfer and the Dissemination of Technological Information within the R&D Organization. Cambridge, Mass: MIT Press.

Allison, Paul David (2001): Missing Data. 1st ed. Thousand Oaks: SAGE Publications (Quantitative Applications in the Social Sciences, 136).

Anderson, J.C; Gerbing, D.W (1988): Structural Equation Modeling in Practice: A Review and Recommended Two-Step Approach. *Psychological Bulletin* 103 (3), pp. 411–423.

Arbuckle, James L. (2006): Amos 7.0 User's Guide. Chicago, Ill: SPSS Inc.

Armitage, Christopher J.; Conner, Mark (1999a): Distinguishing Perceptions of Control from Self-Efficacy: Predicting Consumption of a Low-Fat Diet Using the Theory of Planned Behavior. *Journal of Applied Social Psychology* 29 (1), pp. 72–90.

Armitage, Christopher J.; Conner, Mark (1999b): The Theory of Planned Behaviour: Assessment of Predictive Validity and 'Perceived Control'. *British Journal of Social Psychology* 38 (1), pp. 35–54.

Armitage, Christopher J.; Conner, Mark (2001): Efficacy of the Theory of Planned Behaviour: A Meta-Analytic Review. *British Journal of Social Psychology* 40 (4), pp. 471–499.

Armstrong, J. Scott; Overton, Terry S. (1977): Estimating Nonresponse Bias in Mail Surveys. *Journal of Marketing Research* 14 (3), pp. 396–402.

Arora, Ashish; Fosfuri, Andrea; Gambardella, Alfonso (2001): Markets for Technology and their Implications for Corporate Strategy. *Industrial and Corporate Change* 10 (2), pp. 419–451.

Athanassiou, Nicholas; Nigh, Douglas (2000): Internationalization, Tacit Knowledge and the Top Management Teams of MNCs. *Journal of International Business Studies* 31 (3), p. 471.

Aulakh, Preet S.; Gencturk, Esra F. (2000): International Principal–Agent Relationships: Control, Governance and Performance. *Industrial Marketing Management* 29 (6), pp. 521–538.

Backhaus, Klaus; Erichson, Bernd; Plinke, Wulff; Weiber, Rolf (2011): Multivariate Analysemethoden. Eine anwendungsorientierte Einführung. Extras im Web. 13th ed. Berlin: Springer (Springer-Lehrbuch).

Bagozzi, Richard P. (1980): Causal Models in Marketing. New York, NY, Chichester: Wiley (Theories in Marketing).

Bagozzi, Richard P. (1994): Structural Equation Models in Marketing Research: Basic Principles. In Richard P. Bagozzi (Ed.): Principles of Marketing Research. Cambridge, Mass: Blackwell Business, pp. 317–385.

Bagozzi, Richard P.; Yi, Youjae (1988): On the Evaluation of Structural Equation Models. *Journal of the Academy of Marketing Science* 16 (1), pp. 74–94.

Bailey, James E.; Pearson, Sammy W. (1983): Development of a Tool for Measuring and Analyzing Computer User Satisfaction. *Management Science* 29 (5), pp. 530–545.

Baldwin, Carliss; Hienerth, Christoph; Hippel, Eric von (2006): How User Innovations Become Commercial Products: A Theoretical Investigation and Case Study. *Research Policy* 35 (9), pp. 1291–1313.

Bandura, Albert (1977): Self-Efficacy: Toward a Unifying Theory of Behavioral Change. *Psychological Review* 84 (2), pp. 191–215.

Bandura, Albert (1982): Self-Efficacy Mechanism in Human Agency. *American Psychologist* 37 (2), pp. 122–147.

Bandura, Albert (2003): Self-Efficacy: The Exercise of Control. 8th ed. New York: Freeman.

Barge-Gil, Andrés (2010): Open, Semi-Open and Closed Innovators: Towards an Explanation of Degree of Openness. *Industry and Innovation* 17 (6), pp. 577–607.

Bartol, K. M.; Srivastava, A. (2002): Encouraging Knowledge Sharing: The Role of Organizational Reward Systems. *Journal of Leadership & Organizational Studies* 9 (1), pp. 64–76.

Baughn, C. Christopher; Denekamp, Johannes G.; Stevens, John H.; Osborn, Richard N. (1997): Protecting Intellectual Capital in International Alliances. *Journal of World Business* 32 (2), pp. 103–117.

Bearden, William O.; Netemeyer, Richard G.; Teel, Jesse E. (1989): Measurement of Consumer Susceptibility to Interpersonal Influence. *Journal of Consumer Research* 15 (4), pp. 473–481.

Becerra-Fernandez, Irma; Sabherwal, Rajiv (2010): Knowledge Management: Systems and Processes. Armonk, NY: M.E. Sharpe.

Becker, Jan-Michael; Klein, Kristina; Wetzels, Martin (2012): Hierarchical Latent Variable Models in PLS-SEM: Guidelines for Using Reflective-Formative Type Models. *Long Range Planning* 45 (5-6), pp. 359–394.

Behrends, Thomas (2001): Organisationskultur und Innovativität. Eine kulturtheoretische Analyse des Zusammenhangs zwischen sozialer Handlungsgrammatik und innovativem Organisationsverhalten. Dissertation. München: Hampp (Empirische Personal- und Organisationsforschung, 16).

Berkhout, A. J.; Hartmann, Dap; Duin, Patrick van der; Ortt, Roland (2006): Innovating the Innovation Process. *International Journal of Technology Management* 34 (3-4), pp. 390–404.

Bernard, Harvey Russell (2000): Social Research Methods: Qualitative and Quantitative Approaches. Thousand Oaks, Calif: SAGE Publications.

Blankson, Charles; Kalafatis, Stavros P. (2004): The Development and Validation of a Scale Measuring Consumer/Customer-Derived Generic Typology of Positioning Strategies. *Journal of Marketing Management* 20 (1-2), pp. 5–43.

Blau, Peter Michael (1964): Exchange and Power in Social Life. New York, NY: Wiley.

Blue, Carolyn L. (1995): The Predictive Capacity of the Theory of Reasoned Action and the Theory of Planned Behavior in Exercise Research: An Integrated Literature Review. *Research in Nursing & Health* 18 (2), pp. 105–121.

Bock, Gee-Woo; Kim, Young-Gul (2002): Breaking the Myths of Rewards: An Exploratory Study of Attitudes about Knowledge Sharing. *Information Resources Management Journal* 15 (2), pp. 14–21.

Bock, Gee-Woo; Kim, Young-Gul; Lee, Jae-Nam; Zmud, Robert W. (2005): Behavioral Intention Formation in Knowledge Sharing: Examining the Roles of Extrinsic Motivators, Social-Psychological Forces, and Organizational Climate. *MIS Quarterly* 29 (1), pp. 87–111.

Bogers, Marcel; Afuah, Allan; Bastian, Bettina (2010): Users as Innovators: A Review, Critique, and Future Research Directions. *Journal of Management* 36 (4), pp. 857–875.

Bollen, Kenneth A.; Davis, Walter R. (2009): Causal Indicator Models: Identification, Estimation, and Testing. *Structural Equation Modeling: A Multidisciplinary Journal* 16 (3), pp. 498–522.

Boudreau, Kevin J.; Lakhani, Karim R. (2013): Using the Crowd as an Innovation Partner. *Harvard Business Review* 91 (4), pp. 60–69.

Bradburn, Norman M.; Sudman, Seymour; Wansink, Brian (2004): Asking Questions: The Definitive Guide to Questionnaire Design - For Market Research, Political Polls, and Social and Health Questionnaires. San Francisco, Calif: Jossey-Bass.

Bryman, Alan (2008): Social Research Methods. 3rd ed. Oxford: Oxford University Press.

Bryman, Alan (2010): Mixed Methods in Organizational Research. In David A. Buchanan, Alan Bryman (Eds.): The Sage Handbook of Organizational Research Methods. 2010th ed. Los Angeles: SAGE Publications, pp. 516–531.

Bühl, Achim (2010): PASW 18. Einführung in die moderne Datenanalyse. [ehemals SPSS]. 12th ed. München: Pearson Studium.

Burkhart, Tina; Wuhrmann, Juan; Müller-Kirschbaum, Thomas (2010): Open Innovation und Beziehungsmanagement bei Henkel. *Marketing Review St. Gallen* 27 (4), pp. 14–19.

Cabrera, Elizabeth F.; Cabrera, Ángel (2002): Knowledge-Sharing Dilemmas. *Organization Studies* 23 (4), pp. 683–685.

Caloghirou, Yannis; Kastelli, Ioanna; Tsakanikas, Aggelos (2004): Internal Capabilities and External Knowledge Sources: Complements or Substitutes for Innovative Performance? *Technovation* 24 (1), pp. 29–39.

Cassiman, Bruno; Veugelers, Reinhilde (2006): In Search of Complementarity in Innovation Strategy: Internal R&D and External Knowledge Acquisition. *Management Science* 52 (1), pp. 68–82.

Chandler, Alfred Dupont (1977): The Visible Hand: The Managerial Revolution in American Business. Cambridge, Mass: Belknap Press.

Chan, Johnny; Husted, Kenneth (2010): Dual Allegiance and Knowledge Sharing in Open Source Software Firms. *Creativity and Innovation Management* 19 (3), pp. 314–326.

Chatterji, Deb (1996): Accessing External Sources of Technology. *Research Technology Management* 39 (2), pp. 48–56.

Chatzoglou, Prodromos D.; Vraimaki, Eftichia (2009): Knowledge-Sharing Behaviour of Bank Employees in Greece. *Business Process Management Journal* 15 (2), pp. 245–266.

Chen, Jin; Chen, Yufen; Vanhaverbeke, Wim (2011): The Influence of Scope, Depth, and Orientation of External Technology Sources on the Innovative Performance of Chinese Firms. *Technovation* 31 (8), pp. 362–373.

Chesbrough, Henry William (2003): Open Innovation: The New Imperative for Creating and Profiting from Technology. Boston, Mass: Harvard Business School Press.

Chesbrough, Henry William (2006a): New Puzzles and New Findings. In Henry William Chesbrough, Wim Vanhaverbeke, Joel West (Eds.): Open Innovation. Researching a New Paradigm. Oxford: Oxford University Press, pp. 15–34.

Chesbrough, Henry William (2006b): Open Business Models: How to Thrive in the New Innovation Landscape. Boston, Mass: Harvard Business School Press.

Chesbrough, Henry William (2006c): Open Innovation: A New Paradigm for Understanding Industrial Innovation. In Henry William Chesbrough, Wim Vanhaverbeke, Joel West (Eds.): Open Innovation. Researching a New Paradigm. Oxford: Oxford University Press, pp. 1–12.

Chesbrough, Henry William; Brunswicker, Sabine (2013): Managing Open Innovation in Large Firms. Survey Report – Executive Survey on Open Innovation 2013. Stuttgart: Fraunhofer-Verlag.

Chesbrough, Henry William; Crowther, Adrienne Kardon (2006): Beyond High Tech: Early Adopters of Open Innovation in Other Industries. *R&D Management* 36 (3), pp. 229–236.

Chesbrough, Henry William; Vanhaverbeke, Wim; West, Joel (Eds.) (2006): Open Innovation: Researching a New Paradigm. Oxford: Oxford University Press.

Chiang, Yun-Hwa; Hung, Kuang-Peng (2010): Exploring Open Search Strategies and Perceived Innovation Performance from the Perspective of Inter-Organizational Knowledge Flows. *R&D Management* 40 (3), pp. 292–299.

Child, John and Rodriques Suzana (1996): The Role of Social Identity in the International Transfer of Knowledge through Joint Ventures. In Stewart R. Clegg, Gill Palmer (Eds.): The Politics of Management Knowledge. London: SAGE Publications, pp. 46–68.

Chin, Wynne W. (1998a): Issues and Opinion on Structural Equation Modeling. *MIS Quarterly* 22 (1), pp. 7–16.

Chin, Wynne W. (1998b): The Partial Least Squares Approach to Structural Equation Modeling. In George A. Marcoulides (Ed.): Modern Methods for Business Research. Mahwah, N.J: Lawrence Erlbaum, pp. 295–336.

Chin, Wynne W.; Newsted, Peter R. (1999): Structural Equation Modeling Analysis with Small Samples Using Partial Least Squares. In Rick H. Hoyle (Ed.): Statistical Strategies for Small Sample Research. 2nd ed. Thousand Oaks, Calif: SAGE Publications, pp. 307–342.

Chow, Wing S.; Chan, Lai Sheung (2008): Social Network, Social Trust and Shared Goals in Organizational Knowledge Sharing. *Information & Management* 45 (7), pp. 458–465.

Clagett, Robert Powell (1967): Receptivity to Innovation - Overcoming N.I.H. Master's Thesis. Massachusetts Institute of Technology, Massachusetts. Sloan School of Management. Available online at http://dspace.mit.edu/handle/1721.1/42453.

Cohen, Jacob (1988): Statistical Power Analysis for the Behavioral Sciences. 2nd ed. Hillsdale, NJ: Erlbaum.

Cohen, Wesley M.; Levinthal, Daniel A. (1990): Absorptive Capacity: A New Perspective on Learning and Innovation. *Administrative Science Quarterly* 35 (1), pp. 128–152.

Cole, Jason (2008): How to Deal with Missing Data. Conceptual Overview and Details for Implementing Two Modern Methods. In Jason W. Osborne (Ed.): Best Practices in Quantitative Methods. Los Angeles, Calif: SAGE Publications, pp. 214–238.

Coleman, James Samuel (1990): Foundations of Social Theory. Cambridge, Mass: Belknap Press.

Cong, Xiaoming; Li-Hua, Richard; Stonehouse, George (2007): Knowledge Management in the Chinese Public Sector: Empirical Investigation. *Journal of Technology Management in China* 2 (3), pp. 250–263.

Constant, David; Kiesler, Sara; Sproull, Lee (1994): What's Mine Is Ours, or Is It? A Study of Attitudes about Information Sharing. *Information Systems Research* 5 (4), pp. 400–421.

Cooper, Robert G. (1990): Stage-Gate Systems: A New Tool for Managing New Products. *Business Horizons* 33 (3), pp. 44–54.

Cooper, Robert G. (1996): Overhauling the New Product Process. *Industrial Marketing Management* 25 (6), pp. 465–482.

Cortina, Jose M. (1993): What Is Coefficient Alpha? An Examination of Theory and Applications. *Journal of Applied Psychology* 78, pp. 98–104.

Cronbach, Lee J. (1951): Coefficient Alpha and the Internal Structure of Tests. *Psychometrika* 16 (3), pp. 297-334.

Cummings, Jeffrey L. (2003): Knowledge Sharing: A Review of the Literature. Edited by The World Bank Operations Evaluation Department. The World Bank. Washington D.C. Available online at: http://lnweb90.worldbank.org/oed/oeddoclib.nsf/docunidviewforjavasearch/d9e389e7414be9de85256dc600572ca0/$file/knowledge_eval_li terature_review.pdf.

Cummings, Jeffrey L.; Teng, Bing-Sheng (2003): Transferring R&D Knowledge: The Key Factors Affecting Knowledge Transfer Success. *Journal of Engineering and technology management* 20 (1), pp. 39–68.

Dahlander, Linus; Frederiksen, Lars; Rullani, Francesco (2008): Online Communities and Open Innovation. *Industry and Innovation* 15 (2), pp. 115–123.

Dahlander, Linus; Gann, David M. (2010): How Open Is Innovation? *Research Policy* 39 (6), pp. 699–709.

Dasgupta, Partha; David, Paul A. (1994): Toward a New Economics of Science. *Research Policy* 23 (5), pp. 487–521.

Davenport, Thomas H.; Probst, Gilbert J. B. (Eds.) (2002): Knowledge Management Case Book: Siemens Best Practises. 2nd ed. Erlangen: Publicis-KommunikationsAgentur. Available online at http://www.loc.gov/catdir/description/wiley037/2002284154.html.

Davenport, Thomas H.; Prusak, Laurence (1998): Working Knowledge: How Organizations Manage What They Know. Boston, Mass: Harvard Business School Press.

Deci, Edward L.; Ryan, Richard M. (1985): Intrinsic Motivation and Self-Determination in Human Behavior. New York: Plenum Press (Perspectives in Social Psychology).

DeCuir-Gunby, Jessica T. (2008): Mixed Methods Research in the Social Sciences. In Jason W. Osborne (Ed.): Best Practices in Quantitative Methods. Los Angeles, Calif: SAGE Publications, pp. 125–136.

Diamantopoulos, Adamantios; Riefler, Petra (2011): Using Formative Measures in International Marketing Models: A Cautionary Tale Using Consumer Animosity as an Example. In Marko Sarstedt, Manfred Schwaiger, Charles R. Taylor (Eds.): Measurement and Research Methods in International Marketing. Bingley: Emerald Group Publishing Limited (Advances in International Marketing, 22), pp. 11–30.

Diamantopoulos, Adamantios; Siguaw, Judy A. (2006): Formative Versus Reflective Indicators in Organizational Measure Development: A Comparison and Empirical Illustration. *British Journal of Management* 17 (4), pp. 263–282.

Diamantopoulos, Adamantios; Winklhofer, Heidi M. (2001): Index Construction with Formative Indicators: An Alternative to Scale Development. *Journal of Marketing Research* 38 (2), pp. 269–277.

Dijkstra, Theo (1983): Some Comments on Maximum Likelihood and Partial Least Squares Methods. *Journal of Econometrics* 22 (1–2), pp. 67–90.

Dillman, Don A.; Smyth, Jolene D.; Christian, Leah Melani (2009): Internet, Mail, and Mixed-Mode Surveys. The Tailored Design Method. 3rd ed. Hoboken, NJ: Wiley.

Dixon, Nancy M. (2000): Common Knowledge: How Companies Thrive by Sharing What They Know. Boston, Mass: Harvard Business School Press.

Dodgson, Mark; Gann, David M.; Salter, Ammon (2006): The Role of Technology in the Shift towards Open Innovation: The Case of Procter & Gamble. *R&D Management* 36 (3), pp. 333–346.

Dougherty, Deborah (1992): Interpretive Barriers to Successful Product Innovation in Large Firms. *Organization Science* 3 (2), pp. 179–202.

Drechsler, Wenzel; Natter, Martin (2012): Understanding a Firm's Openness Decisions in Innovation. *Journal of Business Research* 65 (3), pp. 438–445.

Dresing, Thorsten; Pehl, Thorsten (2011): Praxisbuch Transkription. Regelsysteme, Software und praktische Anleitungen für qualitative ForscherInnen. 2nd ed. Marburg: Eigenverlag.

Drucker, Peter F. (1993): Post-Capitalist Society. 1st ed. New York NY: HarperBusiness.

Du Chatenier, Elise; Verstegen, Jos A. A. M.; Biemans, Harm J. A.; Mulder, Martin; Omta, Onno S. W. F. (2010): Identification of Competencies for Professionals in Open Innovation Teams. *R&D Management* 40 (3), pp. 271–280.

Du Preez, Ronel; Visser, Elizabeth; Janse Noordwyk, Hester van (2008): Store Image: Scale Development Part 2. *SA Journal of Industrial Psychology* 34 (2), pp. 50–68.

Dziuban, Charles D.; Shirkey, Edwin C. (1974): When Is a Correlation Matrix Appropriate for Factor Analysis? Some Decision Rules. *Psychological Bulletin* 81 (6), pp. 358–361.

Eckhardt, Andreas; Laumer, Sven; Weitzel, Tim (2009): Who Influences Whom? Analyzing Workplace Referents' Social Influence on IT Adoption and Non-Adoption. *Journal of Information Technology* 24 (1), pp. 11–24.

Edwards, Jeffrey R. (2001): Multidimensional Constructs in Organizational Behavior Research: An Integrative Analytical Framework. *Organizational Research Methods* 4 (2), pp. 144–192.

Eisenberger, Robert; Cameron, Judy (1996): Detrimental Effects of Reward: Reality or Myth? *American Psychologist* 51 (11), pp. 1153–1166.

Eisfeldt, Doreen (2009): Innovatives Arbeitsverhalten Erwerbstätiger. Bestandsaufnahme und wissensbasierte Ansatzpunkte zur Förderung innovativen Arbeitsverhaltens. Dissertation. Hamburg: Kovač (Schriftenreihe Schriften zur Arbeits-, Betriebs- und Organisationspsychologie, 45).

Elmquist, Maria; Fredberg, Tobias; Ollila, Susanne (2009): Exploring the Field of Open Innovation. *European Journal of Innovation Management* 12 (3), pp. 326–345.

Emerson, Richard M. (1976): Social Exchange Theory. *Annual Review of Sociology* 2 (1), pp. 335–362.

Enkel, Ellen (2009): Chancen und Risiken von Open Innovation. In Ansgar Zerfaß, Kathrin M. Möslein (Eds.): Kommunikation als Erfolgsfaktor im Innovationsmanagement. Strategien im Zeitalter der Open Innovation. 1st ed. Wiesbaden: Gabler Verlag, pp. 177–192.

Enkel, Ellen (2010): Attributes Required for Profiting from Open Innovation in Networks. *International Journal of Technology Management* 52 (3), pp. 344–371.

Enkel, Ellen; Bell, John; Hogenkamp, Hannah (2011): Open Innovation Maturity Framework. *International Journal of Innovation Management* 15 (6), pp. 1161–1189.

Enkel, Ellen; Gassmann, Oliver; Chesbrough, Henry William (2009): Open R&D and Open Innovation: Exploring the Phenomenon. *R&D Management* 39 (4), pp. 311–316.

Erden, Zeynep; Krogh, Georg von; Kim, Seonwoo (2012): Knowledge Sharing in an Online Community of Volunteers: The Role of Community Munificence. *European Management Review* 9 (4), pp. 213–227.

Faems, Dries; Looy, Bart van; Debackere, Koenraad (2005): Interorganizational Collaboration and Innovation: Toward a Portfolio Approach. *Journal of Product Innovation Management* 22 (3), pp. 238–250.

Fassott, Georg (2006): Operationalisierung latenter Variablen in Strukturgleichungsmodellen. Eine Standortbestimmung. *Schmalenbachs Zeitschrift für betriebswirtschaftliche Forschung* 58 (1), pp. 67–88.

Felin, Teppo; Foss, Nicolai J. (2005): Strategic Organization: A Field in Search of Micro-Foundations. *Strategic Organization* 3 (4), pp. 441–455.

Feller, Joseph; Fitzgerald, Brian (2002): Understanding Open Source Software Development. London, Boston: Addison-Wesley.

Fengjie, An; Qiao, Fei; Chen, Xin: Knowledge Sharing and Web-based Knowledge-Sharing Platform. *International Conference on E-Commerce Technology for Dynamic E-Business* 2004, pp. 278–281.

Fey, Carl F.; Birkinshaw, Julian (2005): External Sources of Knowledge, Governance Mode, and R&D Performance. *Journal of Management* 31 (4), pp. 597–621.

Fichter, Klaus (2005): Interaktives Innovationsmanagement: Neue Potenziale durch Öffnung des Innovationsprozesses. In Klaus Fichter, Niko Paech, Reinhard Pfriem (Eds.): Nachhaltige Zukunftsmärkte. Orientierungen für unternehmerische Innovationsprozesse im 21. Jahrhundert. Marburg: Metropolis-Verlag (Theorie der Unternehmung, 29), pp. 239–268.

Fishbein, Martin; Ajzen, Icek (1975): Belief, Attitude, Intention and Behavior: An Introduction to Theory and Research. Reading, Mass: Addison-Wesley.

Fleming, Lee; Waguespack, David M. (2007): Brokerage, Boundary Spanning, and Leadership in Open Innovation Communities. *Organization Science* 18 (2), pp. 165–180.

Ford, Dianne P. (2003): Trust and Knowledge Management: The Seeds of Success. In Clyde W. Holsapple (Ed.): Handbook on Knowledge Management. Knowledge Matters. Berlin: Springer (International Handbooks on Information Systems, 1), pp. 353–575.

Fornell, Claes (Ed.) (1982): A Second Generation of Multivariate Analysis. New York: Praeger.

Fornell, Claes (1987): A Second Generation of Multivariate Analysis: Classification of Methods and Implications for Marketing Research. *Review of Marketing* 51, pp. 407–450.

Fornell, Claes; Bookstein, Fred L. (1982): Two Structural Equation Models: LISREL and PLS Applied to Consumer Exit-Voice Theory. *Journal of Marketing Research* 19 (4), pp. 440–452.

Fornell, Claes; Larcker, David F. (1981): Evaluating Structural Equation Models with Unobservable Variables and Measurement Error. *Journal of Marketing Research* 18 (1), pp. 39–50.

Foss, Nicolai J.; Husted, Kenneth; Michailova, Snejina (2010): Governing Knowledge Sharing in Organizations: Levels of Analysis, Governance Mechanisms, and Research Directions. *Journal of Management Studies* 47 (3), pp. 455–482.

Franke, Nikolaus; Hippel, Eric von; Schreier, Martin (2006): Finding Commercially Attractive User Innovations: A Test of Lead-User Theory. *Journal of Product Innovation Management* 23 (4), pp. 301–315.

Gabler Verlag (Ed.) (2004): Gabler Wirtschaftslexikon. Design-Ausgabe zum 75jährigen Verlagsjubiläum. With assistance of Katrin Alisch, Ute Arentzen, Eggert Winter. 16th ed. 4 volumes. Wiesbaden: Gabler Verlag (A-D).

Gardner, Donald G.; Pierce, Jon L. (1998): Self-Esteem and Self-Efficacy within the Organizational Context: An Empirical Examination. *Group & Organization Management* 23 (1), pp. 48–70.

Gassmann, Oliver (2006): Opening Up the Innovation Process: Towards an Agenda. *R&D Management* 36 (3), pp. 223–228.

Gassmann, Oliver; Enkel, Ellen (Eds.) (2004): Towards a Theory of Open Innovation: Three Core Process Archetypes. R&D Management Conference. Lisbon, Portugal, July 6-9.

Gassmann, Oliver; Enkel, Ellen (2006): Open Innovation. Die Öffnung des Innovationsprozesses erhöht das Innovationspotenzial. *Zeitschrift Führung + Organisation* 75 (3), pp. 132–138.

Gassmann, Oliver; Enkel, Ellen; Chesbrough, Henry William (2010): The Future of Open Innovation. *R&D Management* 40 (3), pp. 213–221.

Gavetti, Giovanni (2005): Cognition and Hierarchy: Rethinking the Microfoundations of Capabilities' Development. *Organization Science* 16 (6), pp. 599–617.

Gecas, Viktor (1982): The Self-Concept. *Annual Review of Sociology* 8 (1), pp. 1–33.

Gefen, David; Straub, Detmar W.; Boudreau, Marie-Claude (2000): Structural Equation Modeling and Regression: Guidelines for Research Practice. *Communications of the Association for Information Systems* 4 (7), pp. 1–78.

Geisser, Seymour (1974): A Predictive Approach to the Random Effect Model. *Biometrika* 61 (1), pp. 101–107.

Gemünden, Hans Georg; Salomo, Sören; Hölzle, Katharina (2007): Role Models for Radical Innovations in Times of Open Innovation. *Creativity and Innovation Management* 16 (4), pp. 408–421.

Gemünden, Hans Georg; Walter, Achim (1996): Förderung des Technologietransfers durch Beziehungspromotoren. *Zeitschrift Führung + Organisation* 65 (4), pp. 237–245.

Gibbert, Michael; Krause, Hartmut (2002): Practice Exchange in a Best Practice Marketplace. In Thomas H. Davenport, Gilbert J. B. Probst (Eds.): Knowledge Management Case Book. Siemens Best Practises. 2nd ed. Erlangen: Publicis-KommunikationsAgentur, pp. 89–105.

Gläser, Jochen; Laudel, Grit (2004): Experteninterviews und qualitative Inhaltsanalyse als Instrumente rekonstruierender Untersuchungen. 1st ed. Wiesbaden: VS Verlag für Sozialwissenschaften (Sozialwissenschaften, 2348).

Gordon, Michael E.; Ladd, Robert T. (1990): Dual Allegiance: Renewal, Reconsideration, and Recantation. *Personnel Psychology* 43 (1), pp. 37–69.

Gosling, Samuel D.; Rentfrow, Peter J.; Swann, William B., Jr. (2003): A Very Brief Measure of the Big-Five Personality Domains. *Journal of Research in Personality* 37 (6), pp. 504–528.

Gouldner, Alvin W. (1960): The Norm of Reciprocity: A Preliminary Statement. *American Sociological Review* 25 (2), pp. 161–178.

Graham, John W.; Cumsille, Patricio E.; Shevock, Allison E. (2012): Methods for Handling Missing Data. In John A. Schinka, Wayne F. Verlicer, Irving B. Weiner (Eds.): Handbook of Psychology. Research Methods in Psychology. 2nd ed. Hoboken, NJ: Wiley, pp. 109–141.

Grandori, Anna (2001): Neither Hierarchy nor Identity: Knowledge-Governance Mechanisms and the Theory of the Firm. *Journal of Management & Governance* 5 (3/4), pp. 381–399.

Granovetter, Mark (1985): Economic Action and Social Structure: The Problem of Embeddedness. *American Journal of Sociology* 91 (3), pp. 481–510.

Grant, Robert M. (1996a): Prospering in Dynamically-Competitive Environments: Organizational Capability as Knowledge Integration. *Organization Science* 7 (4), pp. 375–387.

Grant, Robert M. (1996b): Toward a Knowledge-Based Theory of the Firm. *Strategic Management Journal* 17 (Winter Special Issue), pp. 109–122.

Greene, Jennifer C.; Caracelli, Valerie J.; Graham, Wendy F. (1989): Toward a Conceptual Framework for Mixed-Method Evaluation Designs. *Educational Evaluation and Policy Analysis* 11 (3), pp. 255–274.

Groen, Aard J.; Linton, Jonathan D. (2010): Is Open Innovation a Field of Study or a Communication Barrier to Theory Development? *Technovation* 30 (11–12), p. 554.

Grote, Markus (2010): Management geschäftsbereichsübergreifender Innovationsvorhaben. 1st ed. Wiesbaden, Hamburg-Harburg: Gabler Verlag.

Habicht, Hagen; Möslein, Kathrin M.; Reichwald, Ralf (2011): Open Innovation im Unternehmen: Ein Ansatz zur Balance von betrieblichem FuE-Management und Mitarbeiterkreativität. In Dieter Spath (Ed.): Wissensarbeit - Zwischen Strengen Prozessen und Kreativem Spielraum. Berlin: GITO (Schriftenreihe der Hochschulgruppe für Arbeits- und Betriebsorganisation e. V. (HAB), pp. 51–67.

Habicht, Hagen; Möslein, Kathrin M.; Reichwald, Ralf (2012): Open Innovation Maturity. *International Journal of Knowledge-Based Organizations* 2 (1), pp. 92–111.

Hagedoorn, John (1993): Understanding the Rationale of Strategic Technology Partnering: Interorganizational Modes of Cooperation and Sectoral Differences. *Strategic Management Journal* 14 (5), pp. 371–385.

Hagedoorn, John (2002): Inter-Firm R&D Partnerships: An Overview of Major Trends and Patterns Since 1960. *Research Policy* 31 (4), pp. 477–492.

Hagedoorn, John; Duysters, Geert (2002): External Sources of Innovative Capabilities: The Preferences for Strategic Alliances or Mergers and Acquisitions. *Journal of Management Studies* 39 (2), pp. 167–188.

Hair, Joseph F.; Black, William C.; Babin, Barry J.; Anderson, Rolph E. (2008): Multivariate Data Analysis. A Global Perspective. 7th ed. Upper Saddle River, NJ: Pearson.

Hair, Joseph F.; Hult, G. Tomas M.; Ringle, Christian M.; Sarstedt, Marko (2014): A Primer on Partial Least Squares Structural Equations Modeling (PLS-SEM). Thousand Oaks: SAGE Publications.

Hair, Joseph F.; Ringle, Christian M.; Sarstedt, Marko (2011): PLS-SEM: Indeed a Silver Bullet. *Journal of Marketing Theory and Practice* 19 (2), pp. 139–152.

Hair, Joseph F.; Sarstedt, Marko; Ringle, Christian M.; Mena, Jeannette A. (2012): An Assessment of the Use of Partial Least Squares Structural Equation Modeling in Marketing Research. *Journal of the Academy of Marketing Science* 40 (3), pp. 414-433.

Haller, Christine (2003): Verhaltenstheoretischer Ansatz für ein Management von Innovationsprozessen. Universität Stuttgart, Stuttgart. Betriebswirtschaftliches Institut.

Hamel, Gary (1991): Competition for Competence and Interpartner Learning within International Strategic Alliances. *Strategic Management Journal* 12 (S1), pp. 83–103.

Hamel, Gary; Doz, Yves L.; Prahalad, Coimbatore Krishnarao (1989): Collaborate with Your Competitors - and Win. *Harvard Business Review* 67 (1), pp. 133–139.

Han, Kunsoo; Oh, Wonseok; Im, Kun Shin; Oh, Hyelim; Pinsonneault, Alain; Chang, Ray M. (2012): Value Cocreation and Wealth Spillover in Open Innovation Alliances. *MIS Quarterly* 36 (1), pp. 291–316.

Hansen, Morten T. (1999): The Search-Transfer Problem: The Role of Weak Ties in Sharing Knowledge across Organization Subunits. *Administrative Science Quarterly* 44 (1), pp. 82–111.

Hansen, Sean; Avital, Michal (2005): Share and Share Alike: The Social and Technological Influences on Knowledge Sharing Behavior 5 (13).

Harhoff, Dietmar; Henkel, Joachim; Hippel, Eric von (2003): Profiting from Voluntary Information Spillovers. *Research Policy* 32 (10), pp. 1753–1769.

Harkness, S. (2006): Mixed Methods in International Collaborative Research: The Experiences of the International Study of Parents, Children, and Schools. *Cross-Cultural Research* 40 (1), pp. 65–82.

Harrigan, Kathryn R. (1986): Managing for Joint Venture Success. Lexington, Mass: Lexington Books.

Hars, Alexander; Ou, Shaosong (2002): Working for Free? Motivations for Participating in Open-Source Projects. *International Journal of Electronic Commerce* 6, pp. 25–40.

Hauschildt, Jürgen; Salomo, Sören (2011): Innovationsmanagement. 5th ed. München: Vahlen (Vahlens Handbücher der Wirtschafts- und Sozialwissenschaften).

Henkel, Joachim (2003): Software Development in Embedded Linux - Informal Collaboration of Competing Firms. In Wolfgang Uhr, Werner Esswein, Eric Schoop (Eds.): Wirtschaftsinformatik 2003 / Band II. Medien - Märkte - Mobilität: Physica-Verlag, pp. 81-99.

Henkel, Joachim (2006): Selective Revealing in Open Innovation Processes: The Case of Embedded Linux. *Research Policy* 35 (7), pp. 953–969.

Henkel, Joachim (2009): Champions of Revealing: The Role of Open Source Developers in Commercial Firms. *Industrial and Corporate Change* 18 (3), pp. 435–471.

Henseler, Jörg; Ringle, Christian M.; Sarstedt, Marko (2012): Using Partial Least Squares Path Modeling in Advertising Research: Basic Concepts and Recent Issues. In Shintaro Okazaki (Ed.): Handbook of Research on International Advertising. Edward Elgar Publishing, pp. 252–276.

Henseler, Jörg; Ringle, Christian M.; Sinkovics, Rudolf R. (2009): The Use of Partial Least Squares Path Modeling in International Marketing. In Rudolf R. Sinkovics, Pervez N. Ghauri (Eds.): New Challenges to International Marketing. Bradford: Emerald Group Publishing Limited (Advances in International Marketing, 20), pp. 277–319.

Henseler, Jörg; Sarstedt, Marko (2013): Goodness-of-Fit Indices for Partial Least Squares Path Modeling. *Computational Statistics* 28 (2), pp. 565-580.

Herath, Tejaswini; Rao, H. Raghav (2009): Protection Motivation and Deterrence: A Framework for Security Policy Compliance in Organisations. *European Journal of Information Systems* 18 (2), pp. 106–125.

Herstatt, Cornelius; Hippel, Eric von (1992): From Experience: Developing New Product Concepts via the Lead User Method: A Case Study in a "Low-Tech" Field. *Journal of Product Innovation Management* 9 (3), pp. 213–221.

Herstatt, Cornelius; Lüthje, Christian (2005): Quellen für Neuproduktideen. In Sönke Albers, Oliver Gassmann (Eds.): Handbuch Technologie- und Innovationsmanagement. Strategie - Umsetzung - Controlling. 1st ed. Wiesbaden: Gabler Verlag, pp. 265–284.

Herstatt, Cornelius; Nedon, Verena (2014): Open Innovation – Eine Bestandsaufnahme aus Sicht der Forschung und Entwicklung. In Carsten Schultz, Katharina Hölzle (Eds.): Motoren der Innovation – Zukunftsperspektiven der Innovationsforschung. Wiesbaden: Springer Gabler, pp. 247–266.

Herzberg, Frederick (1968): One More Time: How Do You Motivate Employees? *Harvard Business Review* 46 (1), pp. 53–62.

Herzberg, Frederick (1974): Motivation-Hygiene Profiles: Pinpointing What Ails the Organization. *Organizational Dynamics* 3 (2), pp. 18–29.

Herzberg, Frederick; Mausner, Bernard; Snyderman, Barbara Bloch (1959): The Motivation to Work. New York, London: Wiley.

Herzog, Philipp (2008): Open and Closed Innovation: Different Cultures for Different Strategies. Wiesbaden: Gabler Verlag.

Hill, Craig A. (1987): Affiliation Motivation: People Who Need People ... But in Different Ways. *Journal of Personality and Social Psychology* 52 (5), pp. 1008–1018.

Hippel, Eric von (1976): The Dominant Role of Users in the Scientific Instrument Innovation Process. *Research Policy* 5 (3), pp. 212–239.

Hippel, Eric von (1986): Lead Users: A Source of Novel Product Concepts. *Management Science* 32 (7), pp. 791–805.

Hippel, Eric von (1988): The Sources of Innovation. [Repr.]. New York, NY: Oxford University Press.

Hippel, Eric von (1994): "Sticky Information" and the Locus of Problem Solving: Implications for Innovation. *Management Science* 40 (4), pp. 429–439.

Hippel, Eric von (2001): User Toolkits for Innovation. *Journal of Product Innovation Management* 18 (4), pp. 247–257.

Hippel, Eric von (2006): Democratizing Innovation. Cambridge, Mass: MIT Press.

Hippel, Eric von (2010): Comment on 'Is Open Innovation a Field of Study or a Communication Barrier to Theory Development?'. *Technovation* 30 (11–12), p. 555.

Hippel, Eric von; Katz, Ralph (2002): Shifting Innovation to Users via Toolkits. *Management Science* 48 (7), pp. 821–833.

Hippel, Eric von; Krogh, Georg von (2003): Open Source Software and the 'Private-Collective' Innovation Model: Issues for Organization Science. *Organization Science* 14 (2), pp. 209–223.

Hippel, Eric von; Krogh, Georg von (2006): Free Revealing and the Private-Collective Model for Innovation Incentives. *R&D Management* 36 (3), pp. 295–306.

Ho, Chien-Ta Bruce; Hsu, Shih-Feng; Oh, K.B (2009): Knowledge Sharing: Game and Reasoned Action Perspectives. *Industrial Management & Data Systems* 109 (9), pp. 1211–1230.

Hoffmann, Werner H.; Schlosser, Roman (2001): Success Factors of Strategic Alliances in Small and Medium-sized Enterprises – An Empirical Survey. *Long Range Planning* 34 (3), pp. 357–381.

Holmes, Sara; Smart, Palie (2009): Exploring Open Innovation Practice in Firm-Nonprofit Engagements: A Corporate Social Responsibility Perspective. *R&D Management* 39 (4), pp. 394–409.

Holmström, Bengt (1979): Moral Hazard and Observability. *The Bell Journal of Economics* 10 (1), pp. 74–91.

Holmström, Bengt (1982): Moral Hazard in Teams. *The Bell Journal of Economics* 13 (2), pp. 324–340.

Homans, George Caspar (1961): Social Behaviour: Its Elementary Forms. London: Routledge and Kegan Paul.

Howe, Jeff (2006a): Crowdsourcing – Wired Blog Network. Available online at http://crowdsourcing.typepad.com/, checked on 22/01/2014.

Howe, Jeff (2006b): The Rise of Crowdsourcing. *Wired* 14 (6), pp. 176–183.

Howe, Jeff (2009): Crowdsourcing: How the Power of the Crowd is Driving the Future of Business. London: Random House Business.

Hsieh, J. J. Po-An; Rai, Arun; Keil, Mark (2008): Understanding Digital Inequality: Comparing Continued Use Behavioral Models of the Socio-Economically Advantaged and Disadvantaged. *MIS Quarterly* 32 (1), pp. 97–126.

Huang, Qian; Davison, Robert M.; Gu, Jibao (2008): Impact of Personal and Cultural Factors on Knowledge Sharing in China. *Asia Pacific Journal of Management* 25 (3), pp. 451–471.

Huizingh, Eelko (2011): Open Innovation: State of the Art and Future Perspectives. *Technovation* 31 (1), pp. 2–9.

Hulland, John (1999): Use of Partial Least Squares (PLS) in Strategic Management Research: A Review of Four Recent Studies. *Strategic Management Journal* 20 (2), pp. 195–204.

Husted, Kenneth; Michailova, Snejina (2010): Dual Allegiance and Knowledge Sharing in Inter-Firm R&D Collaborations. *Organizational Dynamics* 39 (1), pp. 37–47.

Ili, Serhan; Albers, Albert (2010): Chancen und Risiken von Open Innovation. In Serhan Ili, Albert Albers (Eds.): Open Innovation umsetzen. Prozesse, Methoden, Systeme, Kultur. 1st ed. Düsseldorf: Symposion-Publ, pp. 43–60.

Ipe, Minu (2003): Knowledge Sharing in Organizations: A Conceptual Framework. *Human Resource Development Review* 2 (4), pp. 337–359.

Jaffe, Adam B. (1989): Real Effects of Academic Research. *The American Economic Review* 79 (5), pp. 957–970.

Janzik, Lars (2012): Motivanalyse zu Anwenderinnovationen in Online-Communities. 1st ed. Wiesbaden, Hamburg-Harburg: Gabler Verlag (Gabler Research).

Jarvis, Cheryl Burke; MacKenzie, Scott B.; Podsakoff, Philip M. (2003): A Critical Review of Construct Indicators and Measurement Model Misspecification in Marketing and Consumer Research. *Journal of Consumer Research* 30 (2), pp. 199–218.

Jeon, Suhwan; Kim, Young-Gul; Koh, Joon (2011): An Integrative Model for Knowledge Sharing in Communities-of-Practice. *Journal of Knowledge Management* 15 (2), pp. 251–269.

Jeppesen, Lars Bo (2005): User Toolkits for Innovation: Consumers Support Each Other. *Journal of Product Innovation Management* 22 (4), pp. 347–362.

Jeppesen, Lars Bo; Frederiksen, Lars (2006): Why Do Users Contribute to Firm-Hosted User Communities? The Case of Computer-Controlled Music Instruments. *Organization Science* 17 (1), pp. 45–63.

Jeppesen, Lars Bo; Lakhani, Karim R. (2010): Marginality and Problem-Solving Effectiveness in Broadcast Search. *Organization Science* 21 (5), pp. 1016–1033.

Jewels, Tony; Ford, Marilyn (2006): Factors Influencing Knowledge Sharing in Information Technology Projects. *e-Service Journal* 5 (1), pp. 99–117.

Jöreskog, Karl G. (1970): A General Method for Analysis of Covariance Structures. *Biometrika* 57 (2), pp. 239–251.

Jöreskog, Karl G. (1973): A General Method for Estimating a Linear Structural Equation System. In Arthur Stanley Goldberger, Otis Dudley Duncan (Eds.): Structural Equation Models in the Social Sciences: Seminar Press, pp. 85–112.

Jöreskog, Karl G.; Sörbom, Dag (2001): LISREL 8. User's Reference Guide. 2nd ed., updated to LISREL 8. Lincolnwood: Scientific Software International.

Kaiser, H. F.; Rice, John (1974): Little Jiffy, Mark Iv. *Educational and Psychological Measurement* 34 (1), pp. 111–117.

Kankanhalli, A.; Tan, B.C.Y; Wei, K.K (2005): Contributing Knowledge to Electronic Knowledge Repositories: An Empirical Investigation. *MIS Quarterly*, pp. 113–143.

Karahanna, Elena; Straub, Detmar W.; Chervany, Norman L. (1999): Information Technology Adoption across Time: A Cross-Sectional Comparison of Pre-Adoption and Post-Adoption Beliefs. *MIS Quarterly* 23 (2), pp. 183–213.

Katila, Riitta; Ahuja, Gautam (2002): Something Old, Something New: A Longitudinal Study of Search Behavior and New Product Introduction. *Academy of Management Journal* 45 (6), pp. 1183–1194.

Katz, Ralph; Allen, Thomas J. (1982): Investigating the Not Invented Here (NIH) Syndrome: A Look at the Performance, Tenure, and Communication Patterns of 50 R&D Project Groups. *R&D Management* 12 (1), pp. 7–20.

Kelley, Harold H.; Thibaut, John W. (1978): Interpersonal Relations. A Theory of Interdependence. New York, NY: John Wiley & Sons, Ltd.

Keupp, Marcus Matthias; Gassmann, Oliver (2009): Determinants and Archetype Users of Open Innovation. *R&D Management* 39 (4), pp. 331–341.

Kinch, John W. (1963): A Formalized Theory of the Self-Concept. *The American Journal of Sociology* 68 (4), pp. 481–486.

Kinch, John W. (1973): Social Psychology. New York: McGraw-Hill.

Knudsen, Mette Praest; Mortensen, Thomas Bøtker (2011): Some Immediate – But Negative – Effects of Openness on Product Development Performance. *Technovation* 31 (1), pp. 54–64.

Kock, Ned (2011): Using WarpPLS in e-Collaboration Studies: An Overview of Five Main Analysis Steps. *International Journal of e-Collaboration* 6 (4), pp. 1–13.

Kock, Ned; Chatelain-Jardon, Ruth; Carmona, Jesus (2008): An Experimental Study of Simulated Web-Based Threats and Their Impact on Knowledge Communication Effectiveness. *IEEE Transactions on Professional Communication* 51 (2), pp. 183–197.

Kogut, Bruce (1988): Joint Ventures: Theoretical and Empirical Perspectives. *Strategic Management Journal* 9 (4), pp. 319–332.

Kogut, Bruce; Zander, Udo (1992): Knowledge of the Firm, Combinative Capabilities, and the Replication of Technology. *Organization Science* 3 (3), pp. 383–397.

Kostova, Tatiana (1999): Transnational Transfer of Strategic Organizational Practices: A Contextual Perspective. *Academy of Management Review* 24 (2), pp. 308–324.

Kowal, John; Fortier, Michelle S. (1999): Motivational Determinants of Flow: Contributions from Self-Determination Theory. *Journal of Social Psychology* 139 (3), pp. 355–368.

Krogh, Georg von; Hippel, Eric von (2003): Special Issue on Open Source Software Development. *Research Policy* 32 (7), pp. 1149–1157.

Krogh, Georg von; Ichijō, Kazuo; Nonaka, Ikujirō (2000): Enabling Knowledge Creation: How to Unlock the Mystery of Tacit Knowledge and Release the Power of Innovation. Oxford: Oxford University Press.

Krogh, Georg von; Spaeth, Sebastian; Lakhani, Karim R. (2003): Community, Joining, and Specialization in Open Source Software Innovation: A Case Study. *Research Policy* 32 (7), pp. 1217–1241.

Kuckartz, Udo (2007): Einführung in die computergestützte Analyse qualitativer Daten. 2nd ed. Wiesbaden: VS Verlag für Sozialwissenschaften.

Kuo, Feng-Yang; Young, Mei-Lien (2008a): A Study of the Intention-Action Gap in Knowledge Sharing Practices. *Journal of the American Society for Information Science & Technology* 59 (8), pp. 1224–1237.

Kuo, Feng-Yang; Young, Mei-Lien (2008b): Predicting Knowledge Sharing Practices through Intention: A Test of Competing Models. *Computers in Human Behavior* 24 (6), pp. 2697–2722.

Kwok, Sai Ho; Gao, Sheng (2005): Attitude towards Knowledge Sharing Behavior. *Journal of Computer Information Systems* 46 (2), pp. 45–51.

Lakhani, Karim R. (2008): InnoCentive.com (A). *Harvard Business School Case* (608-170).

Lakhani, Karim R.; Hippel, Eric von (2003): How Open Source Software Works: "Free" User-to-User Assistance. *Research Policy* 32 (6), pp. 923–943.

Lambe, C. Jay; Spekman, Robert E. (1997): Alliances, External Technology Acquisition, and Discontinuous Technological Change. *Journal of Product Innovation Management* 14 (2), pp. 102–116.

Lane, Peter J.; Lubatkin, Michael (1998): Relative Absorptive Capacity and Interorganizational Learning. *Strategic Management Journal* 19 (5), pp. 461–477.

Laursen, Keld; Salter, Ammon (2004): Searching High and Low: What Types of Firms Use Universities as a Source of Innovation? *Research Policy* 33 (8), pp. 1201–1215.

Laursen, Keld; Salter, Ammon (2006): Open for Innovation: The Role of Openness in Explaining Innovation Performance among U.K. Manufacturing Firms. *Strategic Management Journal* 27 (2), pp. 131–150.

Lee, Sungjoo; Park, Gwangman; Yoon, Byungun; Park, Jinwoo (2010): Open Innovation in SMEs: An Intermediated Network Model. *Research Policy* 39 (2), pp. 290–300.

Lee, Yong S. (1996): 'Technology Transfer' and the Research University: A Search for the Boundaries of University-Industry Collaboration. *Research Policy* 25 (6), pp. 843–863.

Lehner, Franz; Haas, Nicolas (2010): Knowledge Management Success Factors - Proposal of an Empirical Research. *Electronic Journal of Knowledge Management* 8 (1), pp. 79–90.

Leonard-Barton, Dorothy (1992): Core Capabilities and Core Rigidities: A Paradox in Managing New Product Development. *Strategic Management Journal* 13 (S1), pp. 111–125.

Lester, Richard K.; McCabe, Mark J. (1993): The Effect of Industrial Structure on Learning by Doing in Nuclear Power Plant Operation. *RAND Journal of Economics* 24 (3), pp. 418–438.

Lettl, Christopher; Herstatt, Cornelius; Gemünden, Hans Georg (2006): Users' Contributions to Radical Innovation: Evidence from Four Cases in the Field of Medical Equipment Technology. *R&D Management* 36 (3), pp. 251–272.

Lichtenthaler, Ulrich (2008): Open Innovation in Practice: An Analysis of Strategic Approaches to Technology Transactions. *IEEE Transactions on Engineering Management* 55 (1), pp. 148–157.

Lichtenthaler, Ulrich (2011): Open Innovation: Past Research, Current Debates, and Future Directions. *Academy of Management* 25 (1), pp. 75–93.

Lichtenthaler, Ulrich; Ernst, Holger (2006): Attitudes to Externally Organising Knowledge Management Tasks: A Review, Reconsideration and Extension of the NIH Syndrome. *R&D Management* 36 (4), pp. 367–386.

Lichtenthaler, Ulrich; Ernst, Holger (2008): Intermediary Services in the Markets for Technology: Organizational Antecedents and Performance Consequences. *Organization Studies* 29 (7), pp. 1003–1035.

Lichtenthaler, Ulrich; Ernst, Holger (2009): Opening up the Innovation Process: The Role of Technology Aggressiveness. *R&D Management* 39 (1), pp. 38–54.

Lichtenthaler, Ulrich; Lichtenthaler, Eckhard (2009): A Capability-based Framework for Open Innovation: Complementing Absorptive Capacity. *Journal of Management Studies* 46 (8), pp. 1315–1338.

Li, Li (2005): The Effects of Trust and Shared Vision on Inward Knowledge Transfer in Subsidiaries' Intra- and Inter-Organizational Relationships. *International Business Review* 14 (1), pp. 77–95.

Limayem, M.; Khalifa, M.; Frini, A. (2000): What Makes Consumers Buy from Internet? A Longitudinal Study of Online Shopping. *Systems, Man and Cybernetics, Part A: Systems and Humans, IEEE Transactions on* 30 (4), pp. 421–432.

Lim, Kwanghui; Chesbrough, Henry William; Ruan, Yi (2010): Open Innovation and Patterns of R&D Competition. *International Journal of Technology Management* 52 (3-4), pp. 295–321.

Lin, Hsiu-Fen (2007a): Effects of Extrinsic and Intrinsic Motivation on Employee Knowledge Sharing Intentions. *Journal of Information Science* 33 (2), pp. 135–149.

Lin, Hsiu-Fen (2007b): Knowledge Sharing and Firm Innovation Capability: An Empirical Study. *International Journal of Manpower* 28 (3/4), pp. 315–332.

Lin, Hsiu-Fen; Lee, Gwo-Guang (2004): Perceptions of Senior Managers toward Knowledge-Sharing Behaviour. *Management Decision* 42 (1), pp. 108–125.

Linstone, Harold A. (2010): Comment on 'Is Open Innovation a Field of Study or a Communication Barrier to Theory Development?'. *Technovation* 30 (11–12), p. 556.

Lippman, Steven A.; Rumelt, Richard P. (1982): Uncertain Imitability: An Analysis of Interfirm Differences in Efficiency Under Competition. *The Bell Journal of Economics* 13 (2), pp. 418–438.

Little, Roderick J. A.; Rubin, Donald B. (1989): The Analysis of Social Science Data with Missing Values. *Sociological Methods & Research* 18 (2-3), pp. 292–326.

Liu, Nien-Chi; Liu, Min-Shi (2011): Human Resource Practices and Individual Knowledge-Sharing Behavior – An Empirical Study for Taiwanese R&D Professionals. *International Journal of Human Resource Management* 22 (4), pp. 981–997.

Li, Ying; Vanhaverbeke, Wim (2009): The Effects of Inter-Industry and Country Difference in Supplier Relationships on Pioneering Innovations. *Technovation* 29 (12), pp. 843–858.

Lohmöller, Jan-Bernd (1989): Latent Variable Path Modeling with Partial Least Squares. Heidelberg: Physica-Verlag.

Lokshin, Boris; Hagedoorn, John; Letterie, Wilko (2011): The Bumpy Road of Technology Partnerships: Understanding Causes and Consequences of Partnership Mal-Functioning. *Research Policy* 40 (2), pp. 297–308.

Lüthje, Christian (2004): Characteristics of Innovating Users in a Consumer Goods Field. *Technovation* 24 (9), pp. 683–695.

Lüthje, Christian; Herstatt, Cornelius (2004): The Lead User Method: An Outline of Empirical Findings and Issues for Future Research. *R&D Management* 34 (5), pp. 553–568.

MacKenzie, Scott B.; Podsakoff, Philip M.; Jarvis, Cheryl Burke (2005): The Problem of Measurement Model Misspecification in Behavioral and Organizational Research and Some Recommended Solutions. *Journal of Applied Psychology* 90 (4), pp. 710–730.

MacKenzie, Scott B.; Podsakoff, Philip M.; Podsakoff, Nathan P. (2011): Construct Measurement and Validation Procedures in MIS and Behavioral Research: Integrating New and Existing Techniques. *MIS Quarterly* 35 (2), pp. 293–334.

Manstead, Antony S. R.; Eekelen, Sander A. M. van (1998): Distinguishing between Perceived Behavioral Control and Self-Efficacy in the Domain of Academic Achievement Intentions and Behaviors. *Journal of Applied Social Psychology* 28 (15), pp. 1375–1392.

Mayring, Philipp (2008): Qualitative Inhaltsanalyse. Grundlagen und Techniken. 10th ed. Beltz: Weinheim.

MaxQDA. Version 10 (2010). Marburg/Berlin: VERBI Software.Consult.Sozialforschung GmbH. Available online at http://www.maxqda.com/.

McQueen, Ronald A.; Knussen, Christina (2002): Research Methods for Social Science: A Practical Introduction. Harlow: Prentice-Hall.

Minbaeva, Dana; Pedersen, Torben (2010): Governing Individual Knowledge-Sharing Behaviour. *International Journal of Strategic Change Management* 2 (2), pp. 200–222.

Miotti, Luis; Sachwald, Frédérique (2003): Co-operative R&D: Why and with Whom? An Integrated Framework of Analysis. *Research Policy* 32 (8), pp. 1481–1499.

Mischke, Johanna; Wingerter, Christian (2012): Frauen und Männer auf dem Arbeitsmarkt. Deutschland und Europa. Edited by Statistisches Bundesamt. Wiesbaden. Available online at https://www.destatis.de/DE/Publikationen/Thematisch/Arbeitsmarkt/Erwerbstaetige/BroeschuereFrauenMaennerArbeitsmarkt.html.

Molm, Linda D. (1997): Coercive Power in Social Exchange. Cambridge, New York: Cambridge University Press (Studies in Rationality and Social Change).

Mortara, Letizia; Thomson, Ruth; Moore, Chris; Armara, Kalliopi; Kerr, Clive; Phaal, Robert; Probert, David (2010): Developing a Technology Intelligence Strategy at Kodak European Research: Scan & Target. *Research Technology Management* 53 (4), pp. 27–38.

Möslein, Kathrin M. (2009): Innovation als Treiber des Unternehmenserfolgs. Herausforderungen im Zeitalter der Open Innovation. In Ansgar Zerfaß, Kathrin M. Möslein (Eds.): Kommunikation als Erfolgsfaktor im Innovationsmanagement. Strategien im Zeitalter der Open Innovation. 1st ed. Wiesbaden: Gabler Verlag, pp. 3–21.

Möslein, Kathrin M.; Bansemir, Bastian (2011): Strategic Open Innovation: Basics, Actors, Tools and Tensions. In Michael Hülsmann, Nicole Pfeffermann (Eds.): Strategies and Communications for Innovations. An Integrative Managment View for Companies and Networks. Berlin, Heidelberg: Springer, pp. 11–24.

Möslein, Kathrin M.; Neyer, Anne-Katrin (2009): Open Innovation. Grundlagen, Herausforderungen, Spannungsfelder. In Ansgar Zerfaß, Kathrin M. Möslein (Eds.): Kommunikation als Erfolgsfaktor im Innovationsmanagement. Strategien im Zeitalter der Open Innovation. 1st ed. Wiesbaden: Gabler Verlag, pp. 85–103.

Mowery, David C.; Oxley, Joanne E.; Silverman, Brian S. (1996): Strategic Alliances and Interfirm Knowledge Transfer. *Strategic Management Journal* 17 (Winter Special Issue), pp. 77–91.

Mullen, Michael R. (1995): Diagnosing Measurement Equivalence in Cross-National Research. *Journal of International Business Studies* 26 (3), pp. 573–596.

Murray, Henry A. (1938): Explorations in Personality. New York, NY: Oxford University Press.

Narula, Rajneesh; Hagedoorn, John (1999): Innovating through Strategic Alliances: Moving towards International Partnerships and Contractual Agreements. *Technovation* 19 (5), pp. 283–294.

Nedon, Verena; Herstatt, Cornelius (2014): R&D Employees' Intention to Exchange Knowledge in Open Innovation Projects. Hamburg University of Technology. Hamburg (Working Paper // Technologie- und Innovationsmanagement, 83). Available online at http://www.tuhh.de/tim/downloads/arbeitspapiere/Working_Paper_83.pdf.

Nelson, Richard R.; Winter, Sidney G. (1982): An Evolutionary Theory of Economic Change. Cambridge, Mass: Belknap Press.

Netemeyer, Richard G.; Sharma, Subhash; Bearden, William O. (2003): Scaling Procedures: Issues and Applications. Thousand Oaks, Calif: SAGE Publications.

Neyer, Anne-Katrin; Bullinger, Angelika C.; Möslein, Kathrin M. (2009): Integrating Inside and Outside Innovators: A Sociotechnical Systems Perspective. *R&D Management* 39 (4), pp. 410–419.

Nicholls-Nixon, Charlene L.; Woo, Carolyn Y. (2003): Technology Sourcing and Output of Established Firms in a Regime of Encompassing Technological Change. *Strategic Management Journal* 24 (7), pp. 651–666.

Nieto, María Jesús; Santamaría, Lluis (2007): The Importance of Diverse Collaborative Networks for the Novelty of Product Innovation. *Technovation* 27 (6–7), pp. 367–377.

Nonaka, Ikujirō; Takeuchi, Hirotaka (1995): The Knowledge Creating Company. How Japanese Companies Create the Dynamics of Innovation. New York, NY: Oxford University Press.

Nonaka, Ikujirō; Toyama, Ryoko; Konno, Noboru (2000): SECI, Ba and Leadership: A Unified Model of Dynamic Knowledge Creation. *Long Range Planning* 33 (1), pp. 5–34.

Nooteboom, Bart (1999): Inter-Firm Alliances: Analysis and Design. London, New York: Routledge.

Nunnally, Jum C.; Bernstein, Ira H. (1994): Psychometric Theory. 3rd ed. New York: McGraw-Hill.

Nuvolari, Alessandro (2004): Collective Invention during the British Industrial Revolution: The Case of the Cornish Pumping Engine. *Cambridge Journal of Economics* 28 (3), pp. 347–363.

Organ, Dennis W.; Konovsky, Mary (1989): Cognitive versus Affective Determinants of Organizational Citizenship Behavior. *Journal of Applied Psychology* 74 (1), pp. 157–164.

Osborne, Jason W. (2008a): Best Practice in Data Transformation. In Jason W. Osborne (Ed.): Best Practices in Quantitative Methods. Los Angeles, Calif: SAGE Publications, pp. 197–204.

Osborne, Jason W. (2008b): Best Practices in Data Cleaning. How Outliers Can Increase Error Rates and Decrease the Quality and Precision of Your Results. In Jason W. Osborne (Ed.): Best Practices in Quantitative Methods. Los Angeles, Calif: SAGE Publications, pp. 205–213.

Osborne, Jason W.; Costello, Anna B.; Kellow, J. Thomas (2008): Best Practice in Exploratory Factor Analysis. In Jason W. Osborne (Ed.): Best Practices in Quantitative Methods. Los Angeles, Calif: SAGE Publications, pp. 86–99.

Osterloh, Margit; Frey, Bruno S. (2000): Motivation, Knowledge Transfer, and Organizational Forms. *Organization Science*, pp. 538–550.

Oxley, Joanne E.; Sampson, Rachelle C. (2004): The Scope and Governance of International R&D Alliances. *Strategic Management Journal* 25 (89), pp. 723–749.

Parkhe, Arvind (1998): Building Trust in International Alliances. *Journal of World Business* 33 (4), pp. 417–437.

PASW Statistics for Windows Version 18.0 (2009). Chicago, Ill: SPSS Inc.

Pavlou, Paul A.; Liang, Huigang; Xue, Yajiong (2007): Understanding and Mitigating Uncertainty in Online Exchange Relationships: A Principal-Agent Perspective. *MIS Quarterly* 31 (1), pp. 105–136.

Pedrosa, Alex Da Mota; Valling, Margus; Boyd, Britta (2013): Knowledge Related Activities in Open Innovation: Managers' Characteristics and Practices. *International Journal of Technology Management* 61 (3-4), pp. 254–273.

Penrose, Edith Tilton (1959): The Theory of the Growth of the Firm. 1st ed. Oxford: Blackwell.

Peteraf, Margaret; Shanley, Mark (1997): Getting to Know You: A Theory of Strategic Group Identity. *Strategic Management Journal* 18 (S1), pp. 165–186.

Peterson, Robert A. (1994): A Meta-Analysis of Cronbach's Coefficient Alpha. *Journal of Consumer Research* 21 (2), pp. 381–391.

Pettersson, Camilla; Lindén-Boström, Margareta; Eriksson, Charli (2009): Reasons for Non-Participation in a Parental Program Concerning Underage Drinking: A Mixed-Method Study. *BMC Public Health* 9 (1), p. 478.

Petter, Stacie; Straub, Detmar W.; Rai, Arun (2007): Specifying Formative Constructs in Information Systems Research. *MIS Quarterly* 31 (4), pp. 623–656.

Pfeffer, Jeffrey; Salancik, Gerald R. (2009): The External Control of Organizations: A Resource Dependence Perspective. Stanford, Calif: Stanford Business Books (Stanford Business Classics).

Pierce, Jon L.; Gardner, Donald G.; Cummings, Larry L.; Dunham, Randall B. (1989): Organization-based Self-Esteem: Construct Definition, Measurement, and Validation. *Academy of Management Journal* 32 (3), pp. 622–648.

Piller, Frank T. (2010): Open Innovation with Customers: Crowdsourcing and Co-Creation at Threadless. SSRN (Working Paper). Available online at http://papers.ssrn.com/sol3/papers.cfm?abstract_id=1688018.

Piller, Frank T.; Möslein, Kathrin M.; Stotko, Christof M. (2004): Does Mass Customization Pay? An Economic Approach to Evaluate Customer Integration. *Production Planning & Control* 15 (4), pp. 435–444.

Piller, Frank T.; Reichwald, Ralf (2009): Wertschöpfungsprinzipien von Open Innovation. Information und Kommunikation in verteilten offenen Netzwerken. In Ansgar Zerfaß, Kathrin M. Möslein (Eds.): Kommunikation als Erfolgsfaktor im Innovationsmanagement. Strategien im Zeitalter der Open Innovation. 1st ed. Wiesbaden: Gabler Verlag, pp. 105–120.

Piller, Frank T.; Walcher, Dominik (2006): Toolkits for Idea Competitions: A Novel Method to Integrate Users in New Product Development. *R&D Management* 36 (3), pp. 307–318.

Pine, B. Joseph (1993): Mass Customization: The New Frontier in Business Competition. Boston, Mass: Harvard Business School Press.

Pirker, Clemens; Füller, Johann; Rieger, Markus; Lenz, Annett (2010): Crowdsoucing im Unternehmensumfeld. In Serhan Ili, Albert Albers (Eds.): Open Innovation umsetzen. Prozesse, Methoden, Systeme, Kultur. 1st ed. Düsseldorf: Symposion-Publ, pp. 315–336.

Plant, Eoin (2009): Modelling Behavioural Antecedents of Inter-Firm Linkages in the Irish Road Freight Industry: An Application of the Theory of Planned Behaviour. Dissertation. Dublin Institute of Technology, Dublin. School of Mechanical and Transport Engineering.

Podsakoff, Philip M.; MacKenzie, Scott B.; Jeong-Yeon Lee; Podsakoff, Nathan P. (2003): Common Method Biases in Behavioral Research: A Critical Review of the Literature and Recommended Remedies. *Journal of Applied Psychology* 88 (5), p. 879.

Podsakoff, Philip M.; Organ, Dennis W. (1986): Self-Reports in Organizational Research: Problems and Prospects. *Journal of Management* 12 (4), p. 531.

Poetz, Marion K.; Schreier, Martin (2012): The Value of Crowdsourcing: Can Users Really Compete with Professionals in Generating New Product Ideas? *Journal of Product Innovation Management* 29 (2), pp. 245–256.

Polanyi, Michael (1966): The Tacit Dimension. 1st ed. Garden City, New York: Doubleday.

Prahalad, Coimbatore Krishnarao; Ramaswamy, Venkat (2004): Co-Creation Experiences: The Next Practice in Value Creation. *Journal of Interactive Marketing* 18 (3), pp. 5–14.

Raasch, Christina; Herstatt, Cornelius; Balka, Kerstin (2009): On the Open Design of Tangible Goods. *R&D Management* 39 (4), pp. 382–393.

Raymond, Eric Steven (1999): The Cathedral and The Bazaar. *Knowledge, Technology & Policy* 12 (3), pp. 23-49.

Reichwald, Ralf; Piller, Frank T. (2009): Interaktive Wertschöpfung. Open Innovation, Individualisierung und neue Formen der Arbeitsteilung. 2nd ed. Wiesbaden: Gabler Verlag.

Reinartz, Werner; Haenlein, Michael; Henseler, Jörg (2009): An Empirical Comparison of the Efficacy of Covariance-based and Variance-based SEM. *International Journal of Research in Marketing* 26 (4), pp. 332–344.

Remneland-Wikhamn, Björn; Ljungberg, J. A. N.; Bergquist, Magnus; Kuschel, Jonas (2011): Open Innovation, Generativity and the Supplier as Peer: The Case of iPhone and Android. *International Journal of Innovation Management* 15 (1), pp. 205–230.

Riege, Andreas (2005): Three-Dozen Knowledge-Sharing Barriers Managers Must Consider. *Journal of Knowledge Management* 9 (3), pp. 18–35.

Rigdon, Edward E. (1998): Structural Equation Modeling. In George A. Marcoulides (Ed.): Modern Methods for Business Research. Mahwah, N.J: Lawrence Erlbaum, pp. 251–294.

Ringle, Christian M.; Götz, Oliver; Wetzels, Martin; Wilson, Bradley (2009): On the Use of Formative Measurement Specifications in Structural Equation Modeling: A Monte Carlo Simulation Study to Compare Covariance-based and Partial Least Squares Model Estimation Methodologies. Edited by Maastricht Research School of Economics of Technology and Organization. Maastricht (Meteor, 09/014).

Ringle, Christian M.; Sarstedt, Marko; Detmar W. Straub (2012): Editor's Comments: A Critical Look at the Use of PLS-SEM in MIS Quarterly. *MIS Quarterly* 36 (1), pp. iii–xiv.

Ringle, Christian M.; Wende, Sven; Will, Alexander (2005): SmartPLS 2.0 (M3) Beta. Hamburg. Available online at http://www. smartpls. de.

Robinson, John P.; Shaver, Philip R.; Wrightsman, Lawrence S. (1991): Criteria for Scale Selection and Evaluation. In John P. Robinson, Philip R. Shaver, Lawrence S. Wrightsman (Eds.): Measures of Personality and Social Psychological Attitudes. San Diego, Calif: Academic Press (Measures of Social Psychological Attitudes, 1), pp. 1–15.

Ryan, Richard M.; Deci, Edward L. (2000): Intrinsic and Extrinsic Motivations: Classic Definitions and New Directions. *Contemporary Educational Psychology* 25 (1), pp. 54–67.

Ryu, Seewon; Ho, Seung Hee; Han, Ingoo (2003): Knowledge Sharing Behavior of Physicians in Hospitals. *Expert Systems with Applications* 25 (1), pp. 113–122.

Sax, Linda J.; Gilmartin, Shannon K.; Bryant, Alyssa N. (2003): Assessing Response Rates and Nonresponse Bias in Web and Paper Surveys. *Research in Higher Education* 44 (4), pp. 409–432.

Schafer, Joseph L.; Graham, John W.; John W. (2002): Missing Data: Our View of the State of the Art. *Psychological Methods* 7 (2), pp. 147–177.

Schattke, Kaspar; Kehr, Hugo M. (2009): Motivation zur Open Innovation. In Ansgar Zerfaß, Kathrin M. Möslein (Eds.): Kommunikation als Erfolgsfaktor im Innovationsmanagement. Strategien im Zeitalter der Open Innovation. 1st ed. Wiesbaden: Gabler Verlag, pp. 121–140.

Schrader, Stephan (1991): Informal Technology Transfer between Firms: Cooperation through Information Trading. *Research Policy* 20 (2), pp. 153–170.

Schreier, Martin; Prügl, Reinhard (2008): Extending Lead-User Theory. *Journal of Product Innovation Management* 25 (4), pp. 331–346.

Schroll, Alexander; Mild, Andreas (2011): Open Innovation Modes and the Role of Internal R&D: An Empirical Study on Open Innovation Adoption in Europe. *European Journal of Innovation Management* 14 (4), pp. 475–495.

Schroll, Alexander; Mild, Andreas (2012): A Critical Review of Empirical Research on Open Innovation Adoption. *Journal für Betriebswirtschaft* 62 (2), pp. 85-118.

Schumpeter, Joseph Alois (1934): The Theory of Economic Development: An Inquiry into Profits, Capital, Credit, Interest, and the Business Cycle. Cambridge, Mass: Harvard Univ. Press (Harvard Economic Studies, 46).

Schüppel, Jürgen: Wissensmanagement. Organisatorisches Lernen im Spannungsfeld von Wissens- und Lernbarrieren. Dissertation, St. Gallen. 1st ed. Wiesbaden: Deutscher Universitäts-Verlag (Gabler Edition Wissenschaft).

Schweisfurth, Tim (2013): Embedded Lead Users inside the Firm: How Innovative User Employees Contribute to the Corporate Product Innovation Process. Wiesbaden: Springer (Forschungs-/ Entwicklungs-/Innovations-Management).

Schweisfurth, Tim; Raasch, Christina (2012): Lead Users as Firm Employees: How Are They Different and Why Does It Matter? SSRN (Academy of Management Annual Meeting). Available online at http://ssrn.com/abstract=2164555.

Schweisfurth, Tim; Raasch, Christina; Herstatt, Cornelius (2011): Free Revealing in Open Innovation: A Comparison of Different Models and their Benefits for Companies. *International Journal of Product Development* 13 (2), pp. 95–118.

Selst, Mark van; Jolicoeur, Pierre (1994): A Solution to the Effect of Sample Size on Outlier Elimination. *The Quarterly Journal of Experimental Psychology Section A* 47 (3), pp. 631–650.

Shook, Christopher L.; Ketchen, David J.; Hult, G. Tomas M.; Kacmar, K. Michele (2004): An Assessment of the Use of Structural Equation Modeling in Strategic Management Research. *Strategic Management Journal* 25 (4), pp. 397–404.

Sieg, Jan Henrik; Wallin, Martin W.; Krogh, Georg von (2010): Managerial Challenges in Open Innovation: A Study of Innovation Intermediation in the Chemical Industry. *R&D Management*, pp. 1–11.

Singh, Jagdip (1995): Measurement Issues in Cross-National Research. *Journal of International Business Studies* 26 (3), pp. 597–619.

Slowinski, Gene (2005): Reinventing Corporate Growth: Implementing the Transformational Growth Model. Gladstone, NJ: Alliance Management Group.

Slowinski, Gene; Sagal, Matthew W. (2010): Good Practice in Open Innovation. *Research Technology Management* 53 (5), pp. 38–45.

Smith, David Horton (1981): Altruism, Volunteers, and Volunteerism. *Nonprofit and Voluntary Sector Quarterly* 10 (1), pp. 21–36.

Snowden, David (2000): The Social Ecology of Knowledge Management. In Charles Despres, Daniele Chauvel (Eds.): Knowledge Horizons. The Present and the Promise of Knowledge Management. Boston: Butterworth-Heinemann, pp. 237–266.

So, Johnny C. F.; Bolloju, Narasimha (2005): Explaining the Intentions to Share and Reuse Knowledge in the Context of IT Service Operations. *Journal of Knowledge Management* 9 (6), pp. 30–41.

Sosik, John J.; Kahai, Surinder S.; Piovoso, Michael J. (2009): Silver Bullet or Voodoo Statistics?: A Primer for Using the Partial Least Squares Data Analytic Technique in Group and Organization Research. *Group & Organization Management* 34 (1), pp. 5–36.

Sparks, Paul; Guthrie, Carol A.; Shepherd, Richard (1997): The Dimensional Structure of the Perceived Behavioral Control Construct. *Journal of Applied Social Psychology* 27 (5), pp. 418–438.

Spender, J.-C (1996): Making Knowledge the Basis of a Dynamic Theory of the Firm. *Strategic Management Journal* 17 (Winter Special Issue), pp. 45–62.

Staw, Barry M. (1976): Intrinsic and Extrinsic Motivation. Morristown, NJ: General Learning Center.

Stone, M. (1974): Cross-Validatory Choice and Assessment of Statistical Predictions. *Journal of the Royal Statistical Society. Series B (Methodological)* 36 (2), pp. 111–147.

Strauss, Anselm L. (1987): Qualitative Analysis for Social Scientists. Cambridge, New York: Cambridge University Press.

Szulanski, Gabriel (1996): Exploring Internal Stickiness: Implementation to the Transfer of Best Practice within the Firm. *Strategic Management Journal* 17 (Winter Special Issue), pp. 27–43.

Tabachnick, Barbara G.; Fidell, Linda S. (2007): Using Multivariate Statistics. 5th ed. Boston, Mass: Pearson/Allyn and Bacon.

Taylor, Joel; Watkinson, David (2007): Indexing Reliability for Condition Survey Data. *Conservator* 30 (1), pp. 49–62.

Teece, David J. (1986): Profiting from Technological Innovation: Implications for Integration, Collaboration, Licensing and Public Policy. *Research Policy* 15 (6), pp. 285–305.

Teece, David J. (2007): Explicating Dynamic Capabilities: The Nature and Microfoundations of (Sustainable) Enterprise Performance. *Strategic Management Journal* 28 (13), pp. 1319–1350.

Teh, Pei-Lee; Ho, Jessica Sze-Yin; Yong, Chen-Chen; Yew, Siew-Yong (2010): Does Internet Self-Efficacy Affect Knowledge Sharing Behavior? In IEEE (Ed.): Proceedings. Industrial Engineering and Engineering Management (IEEM), 2010 IEEE International Conference on Macao, 7.-10.12. Piscataway, NJ, USA; Macao, pp. 94–98.

Teh, Pei-Lee; Yong, Chen-Chen (2011): Knowledge Sharing in IS Personnel: Organizational Behavior's Perspective. *Journal of Computer Information Systems* 51 (4), pp. 11–21.

Temme, Dirk; Kreis, Henning; Hildebrandt, Lutz (2010): A Comparison of Current PLS Path Modeling Software: Features, Ease-of-Use, and Performance. In Vincenzo Esposito Vinzi, Wynne W. Chin, Jörg Henseler, Huiwen Wang (Eds.): Handbook of Partial Least Squares: Concepts, Methods and Applications in Marketing and Related Fields. Berlin: Springer, pp. 737–756.

Tenenhaus, Michel; Amato, Silvano; Vinzi, Vincenzo Esposito (2004): A Global Goodness-of-Fit Index for PLS Structural Equation Modelling. Proceedings of the XLII SIS (Italian Statistical Society) Scientific Meeting, Contributed Papers. Padova, Italy: CLEUP, pp. 739–742.

Tenenhaus, Michel; Vinzi, Vincenzo Esposito; Chatelin, Yves-Marie; Lauro, Carlo (2005): PLS Path Modeling. *Partial Least Squares* 48 (1), pp. 159–205.

Terry, Deborah J.; O'Leary, Joanne E. (1995): The Theory of Planned Behaviour: The Effects of Perceived Behavioural Control and Self-Efficacy. *British Journal of Social Psychology* 34 (2), pp. 199–220.

Tether, Bruce S.; Tajar, Abdelouahid (2008): Beyond Industry–University Links: Sourcing Knowledge for Innovation from Consultants, Private Research Organisations and the Public Science-Base. *Research Policy* 37 (6–7), pp. 1079–1095.

Thomas, David R. (2003): A General Inductive Approach for Qualitative Data Analysis. Edited by New Zealand University of Auckland. School of Population Health (Working Paper).

Tohidinia, Zahra; Mosakhani, Mohammad (2010): Knowledge Sharing Behaviour and Its Predictors. *Industrial Management & Data Systems* 110 (4), pp. 611–631.

Torres, Vasti (2006): A Mixed Method Study Testing Data-Model Fit of a Retention Model for Latino/a Students at Urban Universities. *Journal of College Student Development* 47 (3), pp. 299–318.

Tourangeau, Roger; Rips, Lance J.; Rasinski, Kenneth A. (2000): The Psychology of Survey Response. Cambridge: Cambridge University Press.

Trott, Paul; Hartmann, Dap (2009): Why 'Open Innovation' is Old Wine in New Bottles. *International Journal of Innovation Management* 13 (4), pp. 715–736.

Tschirky, Hugo; Escher, Jean-Philippe; Tokdemir, Deniz; Belz, Christian (2000): Technology Marketing: A New Core Competence of Technology-Intensive Enterprises. *International Journal of Technology Management* 20 (3), pp. 459–474.

Vanhaverbeke, Wim (2006): The Interorganizational Context of Open Innovation. In Henry William Chesbrough, Wim Vanhaverbeke, Joel West (Eds.): Open Innovation. Researching a New Paradigm. Oxford: Oxford University Press, pp. 205–219.

Vanhaverbeke, Wim; Cloodt, Myriam (2006): Open Innnovation in Value Networks. In Henry William Chesbrough, Wim Vanhaverbeke, Joel West (Eds.): Open Innovation. Researching a New Paradigm. Oxford: Oxford University Press, pp. 258–281.

Vanhaverbeke, Wim; Vrande, Vareska van de; Chesbrough, Henry William (2008): Understanding the Advantages of Open Innovation Practices in Corporate Venturing in Terms of Real Options. *Creativity and Innovation Management* 17 (4), pp. 251–258.

Verona, Gianmario; Prandelli, Emanuela; Sawhney, Mohanbir (2006): Innovation and Virtual Environments: Towards Virtual Knowledge Brokers. *Organization Studies* 27 (6), pp. 765–788.

Verworn, Birgit; Herstatt, Cornelius (2000): Modelle des Innovationsprozesses. Hamburg University of Technology. Hamburg (Working Paper // Technologie- und Innovationsmanagement, 6). Available online at http://www.tu-harburg.de/tim/downloads/arbeitspapiere/Arbeitspapier_6.pdf.

Veugelers, Reinhilde (1997): Internal R&D Expenditures and External Technology Sourcing. *Research Policy* 26 (3), pp. 303–315.

Vrande, Vareska van de; Jong, Jeroen P. J. de; Vanhaverbeke, Wim; Rochemont, Maurice de (2009): Open Innovation in SMEs: Trends, Motives and Management Challenges. *Technovation* 29 (6–7), pp. 423–437.

Vrande, Vareska van de; Vanhaverbeke, Wim; Gassmann, Oliver (2010): Broadening the Scope of Open Innovation: Past Research, Current State and Future Directions. *International Journal of Technology Management* 52 (3-4), pp. 221–235.

Walcher, Dominik (2009): Der Ideenwettbewerb als Methode der Open Innovation. In Ansgar Zerfaß, Kathrin M. Möslein (Eds.): Kommunikation als Erfolgsfaktor im Innovationsmanagement. Strategien im Zeitalter der Open Innovation. 1st ed. Wiesbaden: Gabler Verlag, pp. 141–157.

Walliman, Nicholas S. R. (2006): Social Research Methods. London: SAGE Publications (SAGE Course Companions).

Wallin, Martin W.; Krogh, Georg von (2010): Organizing for Open Innovation: Focus on the Integration of Knowledge. *Organizational Dynamics* 39 (2), pp. 145–154.

Wang, S.; Noe, R.A (2010): Knowledge Sharing: A Review and Directions for Future Research. *Human Resource Management Review* 20 (2), pp. 115–131.

Wasko, Molly McLure; Faraj, Samer (2000): It is What One Does: Why People Participate and Help Others in Electronic Communities of Practice. *Journal of Strategic Information Systems* 9 (2-3), pp. 155–173.

Wecht, Christoph H. (2006): Das Management aktiver Kundenintegration in der Frühphase des Innovationsprozesses. Dissertation Universität St. Gallen, 2005. Wiesbaden: Deutscher Universitäts-Verlag (Gabler Edition Wissenschaft).

Weiber, Rolf; Mühlhaus, Daniel (2010): Strukturgleichungsmodellierung. Eine anwendungsorientierte Einführung in die Kausalanalyse mit Hilfe von AMOS, SmartPLS und SPSS. Heidelberg: Springer.

Wernerfelt, Birger (1984): A Resource-Based View of the Firm. *Strategic Management Journal* 5 (2), pp. 171–180.

Wernerfelt, Birger (1995): The Resource-Based View of the Firm: Ten Years After. *Strategic Management Journal* 16 (3), pp. 171–174.

West, Joel (2003): How Open is Open Enough? Melding Proprietary and Open Source Platform Strategies. *Open Source Software Development* 32 (7), pp. 1259–1285.

West, Joel; Bogers, Marcel (2010): Contrasting Innovation Creation and Commercialization within Open, User and Cumulative Innovation (Working Paper). Available online at http://ssrn.com/abstract=1751025.

West, Joel; Bogers, Marcel (2014): Leveraging External Sources of Innovation: A Review of Research on Open Innovation. *Journal of Product Innovation Management* 31 (5), forthcoming. Available online at http://ssrn.com/abstract=2195675

West, Joel; Gallagher, Scott (2006): Challenges of Open Innovation: The Paradox of Firm Investment in Open-Source Software. *R&D Management* 36 (3), pp. 319–331.

West, Joel; Lakhani, Karim R. (2008): Getting Clear About Communities in Open Innovation. *Industry and Innovation* 15 (2), pp. 223–231.

West, Joel; Salter, Ammon; Vanhaverbeke, Wim; Chesbrough, Henry William (2014): Open Innovation: The Next Decade. *Research Policy* 43 (5), pp. 805–811.

West, Joel; Vanhaverbeke, Wim; Chesbrough, Henry William (2006): Open Innovation: A Research Agenda. In Henry William Chesbrough, Wim Vanhaverbeke, Joel West (Eds.): Open Innovation. Researching a New Paradigm. Oxford: Oxford University Press, pp. 285–307.

Wetzels, Martin; Odekerken-Schröder, Gaby; Oppen, Claudia van (2009): Using PLS Path Modeling for Assessing Hierarchical Construct Models: Guidelines and Empirical Illustration. *MIS Quarterly* 33 (1), pp. 177–195.

Whipple, Judith M.; Frankel, Robert (2000): Strategic Alliance Success Factors. *Journal of Supply Chain Management* 36 (2), pp. 21–28.

Winter, Sidney G. (1987): Knowledge and Competence as Strategic Assets. In David J. Teece (Ed.): The Competitive Challenge. Strategies for Industrial Innovation and Renewal. Cambridge, Mass: Ballinger Pub. Co.

Wold, Herman (1966): Nonlinear Estimation by Partial Least Squares Procedures. In Florence Nightingale David (Ed.): Research Papers in Statistics. Festschrift for J. Neyman. New York: Wiley, pp. 411–444.

Wold, Herman (1975): Path Models with Latent Variables: The NIPALS Approach. In Hubert M. Blalock (Ed.): Quantitative Sociology: International Perspectives on Mathematical and Statistical Modeling. New York: Academic Press, pp. 307–357.

Wold, Herman (1982): Soft Modeling: The Basic Design and Some Extensions. In Karl G. Jöreskog, Herman Wold (Eds.): Systems Under Indirect Observation. Causality, Structure, Prediction. Part II. Amsterdam: North-Holland (139), pp. 1–54.

Wu, Hsin-Huan; Wei, Chun-Wang (2010): Factors Affecting Learner's Knowledge Sharing Intentions in Web-based Learning. In IEEE (Ed.): Proceedings. International Symposium on Computer, Communication, Control and Automation (3CA). Tainan, Taiwan, 5.-7.5. IEEE. Piscataway, NJ, USA; Tainan, Taiwan: IEEE (2), pp. 83–86.

Xie, He-Feng (2009): The Determinations of Employee's Knowledge Sharing Behavior: An Empirical Study Based on the Theory of Planned Behavior. In IEEE (Ed.): 16th Annual Conference Proceedings. Management Science and Engineering, 2009. ICMSE 2009. International Conference on. Moscow, Russia, 14.-16.9. IEEE. Piscataway, NJ, USA; Moscow, Russia: IEEE, pp. 1209–1215.

Yang, Heng-Li; Lai, Cheng-Yu (2011): Understanding Knowledge-Sharing Behaviour in Wikipedia. *Behaviour & Information Technology* 30 (1), pp. 131–142.

Zahra, Shaker A.; George, Gerard (2002): Absorptive Capacity: A Review, Reconceptualization, and Extension. *Academy of Management Review* 27 (2), pp. 185–203.

Zaichkowsky, Judith Lynne (1985): Measuring the Involvement Construct. *Journal of Consumer Research*, pp. 341–352.

Zhang, Peihua; Ng, Fung Fai (2012): Attitude toward Knowledge Sharing in Construction Teams. *Industrial Management & Data Systems* 112 (9), pp. 1326–1347

Appendix

Appendix A Interview Guideline

Open Innovation (allgemein)

1. Wann haben Sie das erste Mal etwas von dem Konzept „Open Innovation" gehört?
2. Was verstehen Sie unter Open Innovation ? Wie würde Ihre Definition von 1-2 Sätzen lauten?
3. Was bedeutet Open Innovation für Ihr Unternehmen?
    a. Würden Sie Open Innovation als einen zentralen Bestandteil der Forschung und Entwicklung von Ihrem Unternehmen bezeichnen?
    b. Wie groß ist der Anteil von Open Innovation (Projekten) an der gesamten Forschung Ihres Unternehmen?
4. Inwieweit trägt Open Innovation zum Erfolg Ihres Unternehmens bei?
5. Inwieweit sind Ihre F&E Mitarbeiter mit dem Konzept „Open Innovation" vertraut?
    a. Wird der Begriff „Open Innovation" auch in der internen Kommunikation genutzt?
    b. Können die F&E Mitarbeiter in Ihrer Abteilung mit dem Begriff „Open Innovation" etwas anfangen?
6. Gibt es einen speziellen OI Verantwortlichen in Ihrem Unternehmen?

OI-Projekte (allgemein)

7. Welche Projekte zählen zu OI-Projekten bzw. welche Kriterien entscheiden darüber, ob es sich um ein OI-Projekt handelt?
8. Wie oft haben Sie schon in OI-Projekten gearbeitet?
9. Wie läuft ein OI-Projekt in der Regel in Ihrem Unternehmen ab?
    a. An welchen Stellen ist der Ablauf anders als bei „geschlossenen" Innovationsprojekten?
10. Wo sehen Sie Vorteile von OI-Projekten gegenüber „geschlossenen" Innovationsprojekten, was sind Nachteile?
11. Sind in einem OI-Projekt neben der F&E-Abteilung noch andere Abteilungen an der Zusammenarbeit mit externen Partnern beteiligt?
12. Mit welchen Partnern arbeitet Ihr Unternehmen im Rahmen von OI-Projekten typischerweise zusammen? (Kunden, Wettbewerber, etc.)
    a. Funktioniert die Zusammenarbeit mit allen externen Partnern gleich gut oder gibt es Unterschiede?
    b. Wie kommt man mit den externen Partnern zusammen?
    c. Welche Unterschiede gibt es in der Qualität des Inputs und der Ergebnisse?
    d. Werden unterschiedliche Kommunikationswege genutzt?
13. Wie viele externe Partner sind in der Regel beteiligt?
    a. Wie sieht die Zusammensetzung aus (alles Kunden, Kunden und Supplier, etc.)

Wissensaustausch in OI-Projekten & Rahmenbedingungen

14. Was sind Erfolgsfaktoren für die Zusammenarbeit mit externen Partnern im Rahmen von OI-Projekten?
15. Welche Bedingungen müssen erfüllt sein, damit Ihre Mitarbeiter im Rahmen eines OI-Projekts Wissen mit externen Partnern austauschen?
    a. Welche Dinge behindern oder verhindern sogar den Wissensaustausch mit externen Partnern in OI-Projekten?
16. Gibt es Richt- bzw. Leitlinien, die Ihren Mitarbeitern helfen, in OI-Projekten die richtigen Entscheidungen zu treffen (z.B. im Umgang mit externen Partnern, sensiblen Daten, etc.)?
17. Wissen Ihre Mitarbeiter, welche Informationen Sie im Rahmen eines OI-Projektes nach Außen geben dürfen und welche nicht?
    a. Haben Ihre Mitarbeiter einen Ansprechpartner, wenn sie dennoch mal Fragen bezüglich der Vertraulichkeit und der Weitergabe von Informationen haben?
18. Gibt es spezielle Vereinbarungen/Verträge, die von Ihnen und/oder Ihren Mitarbeitern im Vorfeld eines OI-Projektes unterschrieben und während der Zusammenarbeit mit den externen Partnern beachtet werden müssen?
19. Werden/wurden Ihre F&E-Mitarbeiter in irgendeiner Form auf die Arbeit in OI-Projekten oder auf die Zusammenarbeit mit externen Partnern vorbreitet?
    a. Gab es Schulungen, Gespräche, Anweisungen,...?
20. Gibt es Anreizsysteme oder -mechanismen, die die Arbeit in OI-Projekten oder die Zusammenarbeit mit externen Partnern in OI-Projekten fördern sollen?
    a. Wenn ja, wie sehen diese Anreizsysteme/-mechanismen konkret aus?
21. Was für Befürchtungen und Ängste existieren bei Ihren Mitarbeitern bzgl. des Wissensaustauschs mit externen Partnern in OI-Projekten?
22. Arbeiten Ihre Mitarbeiter lieber in OI-Projekten oder in geschlossenen Innovationsprojekten?

Appendix B Questionnaire (Limesurvey)

Open Innovation (OI) and why R&D employees participate

English

Thank you very much for participating in this survey about Open Innovation.

In the course of my dissertation at the Hamburg University of Technology I investigate why R&D employees exchange knowledge with external partners in Open Innovation projects and what may hinder them, respectively. With your participation you do not only support my research project to a very high degree, but also help (your) company to understand what needs to be done so that you are able and willing to participate in Open Innovation projects.

In order to assure a common understanding of three key words of the survey, please find a short explanation below:

<u>Open Innovation project (in short: OI-project)</u> = innovation project, where your company cooperates with one or more external partners. Thereby, own (internal) ideas/technologies are combined with external ideas/technologies in order to accelerate and/or advance the own innovation process.

<u>External partner</u> = person/institution that is considered as outside the company boundaries, e.g., customer, supplier, universities etc. (business units or subsidiaries of the own company are no external partners)

<u>Knowledge exchange</u> = reciprocal knowledge sharing process, where both/all parties receive knowledge as well as share knowledge.

Responding all questions of the survey will take **7-10 minutes**. The survey is **anonymous** and the collected data will be used **exclusively for the purpose of my thesis research**. If you have any questions or comments, do not hesitate to contact me (verena.nedon@tuhh.de).

TUHH
Technische Universität Hamburg-Harburg

[Exit and clear survey]   [Load unfinished survey]   [Next >>]

Open Innovation (OI) and why R&D employees participate

0% [================] 100%

English ▼

1 How often did you work together in Open Innovation projects (OI-projects) with the following partners?

	very rarely	rarely	occasionally	often	very often
Universities / Research Instituts	○	○	○	○	○
Customers	○	○	○	○	○
Suppliers	○	○	○	○	○
Competitors	○	○	○	○	○
Other industrial partners	○	○	○	○	○

2 In how many OI-projects did you approximately work?

Only numbers may be entered in these fields

During the last 3 years: [____] # Projekte / projects
During the last 10 years: [____] # Projekte / projects

3 Which requirements must be met, so that you are able to exchange knowledge with external partners in OI-projects?

a) [_____]
b) [_____]
c) [_____]
d) [_____]
e) [_____]

[Exit and clear survey]   [Resume later]   [Next >>]

Open Innovation (OI) and why R&D employees participate

* 4 To what extent do you agree or disagree with the following statements: My knowledge exchange with external partners in OI-projects...

	strongly disagree				strongly agree
...helps other members in my organization to solve problems.	○	○	○	○	○
...creates new business opportunities for my organization.	○	○	○	○	○
...improves work processes in my organization.	○	○	○	○	○
...increases productivity in my organization.	○	○	○	○	○
...helps my organization to achieve its performance objectives.	○	○	○	○	○
...is part of my job.	○	○	○	○	○

* 5 To what extent do you agree or disagree with the following statements: When I exchange knowledge with external partners in OI-projects it is important to me...

	strongly disagree				strongly agree
...to get better work assignments.	○	○	○	○	○
...to be promoted.	○	○	○	○	○
...to get a higher salary.	○	○	○	○	○
...to get a higher bonus.	○	○	○	○	○
...to increase my job security.	○	○	○	○	○
...to enhance my reputation.	○	○	○	○	○
...to build a network.	○	○	○	○	○
...to increase my knowledge.	○	○	○	○	○
...to improve my job performance.	○	○	○	○	○
...to add value for my company.	○	○	○	○	○

* 6 To what extent do you agree or disagree with the following statements: When I exchange knowledge with external partners in OI-projects...

	strongly disagree				strongly agree
...I believe that I will get knowledge for giving knowledge.	○	○	○	○	○
...I expect somebody to respond when I'm in need.	○	○	○	○	○
...I expect to get back knowledge when I need it.	○	○	○	○	○
...I believe that my queries for knowledge will be answered in future.	○	○	○	○	○

* 7 To what extent do you agree or disagree with the following statements:

	strongly disagree				strongly agree
I enjoy exchanging knowledge with external partners in OI-projects.	○	○	○	○	○
I enjoy helping others by exchanging knowledge with external partners in OI-projects.	○	○	○	○	○
It feels good to help someone else by exchanging knowledge with external partners in OI-projects.	○	○	○	○	○

[Exit and clear survey]   [Resume later]   [Next >>]

Open Innovation (OI) and why R&D employees participate

0% [====   ] 100%

English ▼

8 My knowledge exchange with external partners in OI-projects is...
Choose one of the following answers

○ very harmful
○ harmful
○ neither harmful nor beneficial
○ beneficial
○ very beneficial

9 My knowledge exchange with external partners in OI-projects is a ... experience.
Choose one of the following answers

○ very unpleasant
○ unpleasant
○ neither unpleasant nor pleasant
○ pleasant
○ very pleasant

10 My knowledge exchange with external partners in OI-projects is ... to me.
Choose one of the following answers

○ very worthless
○ worthless
○ neither worthless nor valuable
○ valuable
○ very valuable

11 My knowledge exchange with external partners in OI-projects is a ... move.
Choose one of the following answers

○ very unwise
○ unwise
○ neither unwise nor wise
○ wise
○ very wise

12 Overall, my knowledge exchange with external partners in OI-projects is...
Choose one of the following answers

○ very bad
○ bad
○ neither bad nor good
○ good
○ very good

[Exit and clear survey]   [Resume later]   [Next >>]

Open Innovation (OI) and why R&D employees participate

13 To what extent do you consider the following statements as likely or unlikely: I will exchange...

	very unlikely				very likely
...work reports and official documents with external partners in future OI-projects.	○	○	○	○	○
...manuals, methodologies and models with external partners in future OI-projects.	○	○	○	○	○
...experience or know-how from work with external partners in future OI-projects.	○	○	○	○	○
...know-where or know-whom at the request of external partners in OI-projects.	○	○	○	○	○
...my expertise from my education or training with external partners in future OI-projects.	○	○	○	○	○

14 To what extent do you consider the following statements as likely or unlikely:

	very unlikely				very likely
My CEO wants me to exchange knowledge with external partners in OI-projects.	○	○	○	○	○
My immediate supervisor wants me to exchange knowledge with external partners in OI-projects.	○	○	○	○	○
My colleagues want me to exchange knowledge with external partners in OI-projects.	○	○	○	○	○

15 To what extent do you agree or disagree with the following statements: Generally speaking...

	strongly disagree				strongly agree
...I try to follow the CEO's policy and intention.	○	○	○	○	○
...I accept and carry out my immediate supervisor's decision even though it is different from mine.	○	○	○	○	○
...I respect and put in practice my colleagues' decision.	○	○	○	○	○

16 To what extent do you agree or disagree with the following statements:

	strongly disagree				strongly agree
Whether or not I exchange knowledge with external partners in OI-projects is entirely up to me.*	○	○	○	○	○
I have full personal control over exchanging knowledge with external partners in OI-projects.*	○	○	○	○	○
If it is entirely up to me, I am confident that I am able to exchange knowledge with external partners in OI-projects.*	○	○	○	○	○
I believe I have the ability to exchange knowledge with external partners in OI-projects.*	○	○	○	○	○
I am capable of exchanging knowledge with external partners in OI-projects.*	○	○	○	○	○

* within the contractual agreements

17 For me exchanging knowledge with external partners in OI-projects is...
Choose one of the following answers

○ very difficult
○ difficult
○ neither difficult nor easy
○ easy
○ very easy

Open Innovation (OI) and why R&D employees participate

0% ▬▬▬▬▬▬ 100%

English

18 To what extent do you agree or disagree with the following statements: I see myself as...

	strongly disagree				strongly agree
... extroverted, enthusiastic.	○	○	○	○	○
... critical, quarrelsome.	○	○	○	○	○
... dependable, self-disciplined.	○	○	○	○	○
... anxious, easily upset.	○	○	○	○	○
... open to new experiences, complex.	○	○	○	○	○
... reserved, quiet.	○	○	○	○	○
... sympathetic, warm.	○	○	○	○	○
... disorganized, careless.	○	○	○	○	○
... calm, emotionally stable.	○	○	○	○	○
... conventional, uncreative.	○	○	○	○	○

19 How old are you?

____ Jahre / years

Only numbers may be entered in this field

20 Please state your gender.

○ Female ○ Male ● No answer

21 Please state your highest level of education completed.
Choose one of the following answers

[Please choose... ▼]

22 In which field of expertise did you earn your highest educational degree? *(e.g. chemical physics, material science, business administration etc.)*

_____

23 How long do you already work for your company?

____ Jahre / years

Only numbers may be entered in this field

24 Which country do you work in?

_____

25 Comments / Feedback

[   ]

❓ You have completed the survey. If you have any comments or feedback on the general topic or the questionnaire, please feel free to use the free text field.

Even if you do not have any comments, please press "Submit" to save your answers.

Thank you very much for taking the time and completing my survey.

Appendix C Results of Confirmatory Factor Analysis (CFA) with All Items

		INDICATOR RELIABILITY		INTERNAL CONSISTENCY RELIABILITY		CONVERGENT VALIDITY
		Standardized Indicator Loading λ	T-Value	Dillon-Goldstein's ρ	Standardized Cronbach's α	Average Variance Extracted
Critical Value		λ ≥ 0.7	≥ 1.96: p<0.05 ≥ 2.58: p<0.01 ≥ 3.29: p<0.001	ρ ≥ 0.7	0.7 ≤ α ≤ 0.9	AVE ≥ 0.5
Construct	Item					
Attitude	A1	0.650	5.436	0.813	0.714	0.466
	A2	0.666	8.992			
	A3	0.741	14.833			
	A4	0.635	8.272			
	A5	0.716	12.463			
Perceived Behavioral Control	PBC1	0.714	13.476	0.885	0.843	0.566
	PBC2	0.765	17.387			
	PBC3	0.786	19.336			
	PBC4	0.837	19.038			
	PBC5	0.815	18.285			
	PBC6	0.568	7.810			
Intention[175] (2nd order)	Path_1	0.800	19.946	0.867	n.a.	0.748
	Path_2	0.924	66.630			
Intention_doc (1st order)	I1	0.857	27.080	0.862	0.681	0.758
	I2	0.884	45.631			
Intention_undoc (1st order)	I3	0.816	20.651	0.888	0.811	0.727
	I4	0.861	27.784			
	I5	0.879	37.272			
Enjoyment in Helping	JOY1	0.856	24.163	0.887	0.814	0.725
	JOY2	0.926	49.479			
	JOY3	0.763	9.683			
Sense of Self-Worth	SW1	0.705	11.880	0.830	0.752	0.497
	SW2	0.642	8.184			
	SW3	0.599	6.946			
	SW4	0.776	14.211			
	SW5	0.785	13.999			
Reciprocity	RP1	0.770	12.188	0.847	0.773	0.582
	RP2	0.703	6.705			
	RP3	0.773	7.824			
	RP4	0.802	13.856			

---

[175] Path 1 represents the relation between the second-order construct intention and the first-order construct intention_doc. Path 2 represents the relation between intention and intention_undoc. In case of second-order constructs, Cronbach's alpha is only calculated for the first-order constructs (see Wetzels et al. 2009).

		INDICATOR RELIABILITY		INTERNAL CONSISTENCY RELIABILITY		CONVERGENT VALIDITY
		Standardized Indicator Loading λ	T-Value	Dillon-Goldstein's ρ	Standardized Cronbach's α	Average Variance Extracted
Critical Value		λ ≥ 0.7	≥ 1.96: p<0.05 ≥ 2.58: p<0.01 ≥ 3.29: p<0.001	ρ ≥ 0.7	0.7 ≤ α ≤ 0.9	AVE ≥ 0.5
Construct	**Item**					
Reward A	REW1	0.757	3.312	0.899	0.870	0.599
	REW2	0.824	3.207			
	REW3	0.819	3.030			
	REW4	0.826	3.143			
	REW5	0.662	2.817			
	REW6	0.742	3.185			
Reward B	REW7	0.735	11.181	0.827	0.720	0.547
	REW8	0.866	20.791			
	REW9	0.617	6.757			
	REW10	0.719	10.226			

Bootstrapping conducted with 133 cases and 8,000 samples

Appendix D MSA, Communalities and Pattern Matrix of Remaining Items

	MSA	Comm-unality	Component								
			1	2	3	4	5	6	7	8	9
Critical value	> 0.5	> 0.4	Loading > 0.5								
A2	0.779	0.597	**0.519**	-0.024	-0.399	0.090	0.149	0.054	-0.032	-0.034	0.235
A3	0.891	0.583	**0.654**	0.017	0.085	-0.051	-0.021	0.060	-0.076	0.073	0.190
A4	0.750	0.664	**0.737**	0.210	-0.056	-0.205	0.119	0.105	0.178	-0.041	-0.344
A5	0.770	0.651	**0.799**	-0.235	0.177	0.186	0.037	-0.219	-0.012	0.002	0.009
PBC1	0.663	0.709	0.039	**0.736**	-0.263	0.125	-0.186	-0.063	0.141	0.009	0.071
PBC2	0.723	0.734	0.212	**0.724**	-0.114	0.199	-0.164	-0.103	-0.087	0.131	0.017
PBC3	0.864	0.667	-0.163	**0.791**	0.128	0.165	0.058	0.047	-0.030	0.111	-0.042
PBC4	0.758	0.829	-0.082	**0.776**	0.241	-0.147	0.157	-0.042	-0.046	-0.113	0.169
PBC5	0.738	0.814	-0.033	**0.740**	0.308	-0.156	0.069	-0.006	-0.012	-0.112	0.127
I1	0.781	0.689	0.041	0.098	**0.779**	0.385	-0.022	-0.132	0.098	0.055	-0.187
I2	0.726	0.627	0.184	0.007	**0.578**	0.537	0.096	0.025	-0.072	0.002	-0.244
I3	0.820	0.654	-0.053	0.229	0.273	**0.680**	0.098	0.057	-0.038	0.106	-0.091
I4	0.808	0.687	0.029	-0.031	0.106	**0.740**	-0.085	0.051	-0.005	-0.091	0.217
I5	0.813	0.747	-0.039	0.003	0.184	**0.806**	0.019	0.002	0.003	-0.121	0.113
JOY1	0.799	0.681	0.178	0.076	-0.007	-0.032	**0.720**	0.046	-0.124	-0.076	0.162
JOY2	0.693	0.827	0.141	-0.061	-0.024	0.035	**0.821**	0.047	-0.010	-0.027	0.078
JOY3	0.651	0.693	-0.104	-0.023	-0.048	0.042	**0.777**	0.057	0.112	0.068	-0.005
SW1	0.812	0.546	-0.068	0.030	-0.132	0.119	0.094	**0.577**	-0.027	-0.075	0.300
SW3	0.663	0.674	-0.140	-0.131	-0.092	0.126	0.062	**0.808**	-0.121	0.105	-0.060
SW4	0.653	0.759	0.010	0.000	-0.051	-0.051	0.108	**0.870**	0.061	0.044	-0.164
SW5	0.631	0.666	0.214	0.005	0.267	-0.086	-0.165	**0.577**	0.053	0.051	0.145
RP2	0.686	0.758	0.074	0.015	0.054	-0.165	-0.002	-0.034	**0.850**	0.070	-0.005
RP3	0.645	0.822	0.013	-0.062	0.174	-0.027	-0.002	-0.085	**0.880**	-0.053	0.065
RP4	0.819	0.690	-0.058	0.026	-0.108	0.250	0.019	0.129	**0.715**	-0.047	0.024
REW1	0.675	0.528	-0.069	-0.075	0.089	0.043	0.288	0.042	0.153	**0.543**	-0.024
REW2	0.821	0.850	0.077	-0.058	-0.003	-0.007	-0.053	0.073	-0.011	**0.910**	-0.017
REW3	0.696	0.872	-0.028	0.072	0.019	-0.061	-0.012	0.058	-0.067	**0.955**	-0.006
REW4	0.726	0.875	0.009	0.091	0.011	-0.064	-0.099	0.084	-0.022	**0.953**	0.019
REW6	0.821	0.671	-0.013	-0.083	0.041	-0.017	0.130	-0.320	0.072	**0.552**	0.466
REW7	0.693	0.644	0.041	0.123	-0.243	-0.018	0.087	-0.055	-0.057	0.065	**0.768**
REW8	0.817	0.746	-0.109	0.106	-0.070	0.118	0.126	0.000	0.113	-0.024	**0.751**
REW10	0.787	0.674	0.065	-0.050	0.332	0.079	-0.241	0.279	0.091	-0.018	**0.498**

Extraction method: principal component analysis
Rotation method: promax with Kaiser-normalization

MIX
Papier aus verantwortungsvollen Quellen
Paper from responsible sources
FSC® C105338

If you have any concerns about our products,
you can contact us on
ProductSafety@springernature.com

In case Publisher is established outside the EU,
the EU authorized representative is:
**Springer Nature Customer Service Center GmbH
Europaplatz 3, 69115 Heidelberg, Germany**

Printed by Libri Plureos GmbH
in Hamburg, Germany